PHYSICAL HAZARD CONTROL
Preventing Injuries in the Workplace

Frank R. Spellman and Revonna M. Bieber

GOVERNMENT INSTITUTES

An imprint of
THE SCARECROW PRESS, INC.
Lanham • Toronto • Plymouth, UK
2011

**Government
Institutes**

Published by Government Institutes
An imprint of The Scarecrow Press, Inc.
A wholly owned subsidiary of The Rowman & Littlefield Publishing Group, Inc.
4501 Forbes Boulevard, Suite 200, Lanham, Maryland 20706
http://www.govinstpress.com

Estover Road, Plymouth PL6 7PY, United Kingdom

Copyright © 2011 by Government Institutes

Photographs by Danielle O'Pezio

British Library Cataloguing in Publication Information Available

Library of Congress Cataloging-in-Publication Data
Spellman, Frank R.
  Physical hazard control : preventing injuries in the workplace / Frank R. Spellman and Revonna M. Bieber.
    p. cm.
  Includes bibliographical references and index.
  ISBN 978-1-60590-761-1 (hardback) — ISBN 978-1-60590-762-8 (electronic)
  1. Industrial safety. 2. Industrial accidents—Prevention. I. Bieber, Revonna M., 1976–
II. Title.
  T55.S6447 2011
  363.11—dc23
                                                                        2011021851

∞™ The paper used in this publication meets the minimum requirements of American National Standard for Information Sciences—Permanence of Paper for Printed Library Materials, ANSI/NISO Z39.48-1992.

Printed in the United States of America

Contents

Preface   v

List of Acronyms and Abbreviations   vii

1   Physical Hazards and Controls   1
2   Layout, Construction, and Maintenance of Facilities   8
3   Safeguarding, Lockouts, and Tagouts   20
4   Confined Spaces   28
5   Noise   40
6   Radiation   47
7   Ergonomics   54
8   Electrical Safety   59
9   Fire Safety and Thermal Stressors   66
10   Hand and Portable Power Tools   78
11   Woodworking   81
12   Metalworking Machinery   89
13   Welding Operations   96
14   Vehicular and Mobile Equipment Safety   101
15   Retail, Service, and Warehouse Facilities   111
16   Workplace Violence   127
17   Safety Evaluations and Inspection Processes   134

References and Recommended Reading   143

Index   147

About the Authors   151

Preface

PEOPLE DEAL WITH HAZARDS EVERYDAY at the workplace, in their homes, and on the roadways; the list goes on and on. Within each and every situation, people are faced with potential hazards. The main focus of this book deals with physical hazards in the workplace and the controls to prevent injury, illness, and death. A variety of physical dangers can be present in any one operation. Physical hazards stressed throughout the text include layout and building design, safeguarding of machinery, confined space entry, noise, radiation, ergonomics, electricity, thermal stressors, hand-tools, woodworking, welding, machining, mobile equipment, materials handling, and workplace violence. Engineering controls, administrative controls (to include safe work practices), and the use of personal protective equipment (PPE) are discussed for each physical hazard presented, along with real-world examples and solutions.

This book was written as an up-to-date practical guide focusing on a variety of physical hazards and controls. It can be used as a textbook for safety students, a quick reference for safety professionals, a refresher for those preparing to take the certified safety professionals (CSP) exam or the certified industrial hygienists (CIH) exam, or a guide for anyone needing information regarding physical hazards present in industry, maintenance, and the service industry.

Acronyms and Abbreviations

ABS	American Bureau of Shipping
ADA	Americans with Disabilities Act
AGV	automated guided vehicle
AL	action limit
ALARA	as low as reasonably achievable
ANSI	American National Standards Institute
ASME	American Society of Mechanical Engineers
ASSE	American Society of Safety Engineers
ASTM	American Society for Testing Materials
BLS	Bureau of Labor Statistics
CDC	Centers for Disease Control
CDL	commercial driver's license
CIH	certified industrial hygienist
CPM	computerized predictive maintenance
CSP	certified safety professional
CTD	cumulative trauma disorder
dB	decibels
DHHS	Department of Health and Human Services
DOT	Department of Transportation
EEO	equal employment opportunity
ELF	extremely low frequency
EMS	emergency medical service
EPA	Environmental Protection Agency
FAA	Federal Aviation Administration
FMCSR	Federal Motor Carrier Safety Regulations

GFIC	ground fault interrupter circuit
GHz	gigahertz
HEI	House Ear Institute
HEPA	high-efficiency particulate air
HP	hypersensitivity pneumonitis
Hz	hertz
IDLH	immediately dangerous to life and health
IR	infrared
JA	job analysis
JHA	job hazard analysis
kHz	kilohertz
laser	light amplification by stimulated emission of radiation
LOTO	lockout/tagout
LPG	liquefied petroleum gas
LRV	light-reflectance value
MHz	megahertz
MPL	maximum permissible lift
MSD	musculoskeletal disorder
MSHA	Mine Safety and Health Administration
MW	microwaves
MWF	metalworking fluid
NB	National Board of Boiler and Pressure Vessel
NEC	National Electrical Code
NFPA	National Fire Protection Association
NIHL	noise-induced hearing loss
NIOSH	National Institute for Occupational Safety and Health
NRR	Noise Reduction Rating
OSH	Occupational Safety and Health
OSHA	Occupational Safety and Health Administration
PAH	polynuclear aromatic hydrocarbon
PIT	powered industrial truck
PPE	personal protective equipment
RF	radio frequency
SCBA	self-contained breathing apparatus
SHARE	Safety, Health, and Return-to-Employment
SMAW	shielded metal arc welding
SOLAS	safety of life at sea
STS	standard threshold shift
TWA	time-weighted average
UAW	United Auto Workers
USCG	U.S. Coast Guard
USDOE	U.S. Department of Energy
UV	ultraviolet light
WMSD	work-related musculoskeletal disorder

1

Physical Hazards and Controls

W E DEAL WITH EXPOSURE TO HAZARDS as a "condition of employment." Employees are exposed to these hazards because they are required to work in environments with certain machines, tools, or chemicals. For instance, consumers using a power drill take personal responsibility for their safety, but for an employee using that same drill provided by an employer and required to be used on the job, the employer assumes responsibility for the worker's safety.

What Is a Hazard?

A hazard can be defined as a condition with the potential for causing injury to personnel, damage to equipment or structures, loss of material, or lessening of the ability to perform a prescribed function. When a hazard is present, there is a possibility of one or more of these adverse effects occurring.

There are basically three types of hazards one may encounter in the workplace: physical hazards, chemical hazards, and biological hazards. This chapter focuses on physical hazards, which are categorized as:

- inherent properties or characteristics of the equipment, ranging from high-voltage equipment to the sharp blade of a saw

- failures, such as failure of a safeguard to function, a pressure vessel gauge blowing off because the vessel is overheated, human failure to follow a complex procedure, or unsafe acts
- environmental stresses, such as rain that causes metal corrosion on a platform or ladder, lightning strikes causing voltage overloads, or high humidity and heat causing personal stress (Wald and Stave, 2002).

Each product, system, or operation is studied for the existence or potential of any of these three physical hazard categories. Hazards can be present, but if they are under suitable control, they present little danger. They are instances where hazards cannot be eliminated, so they must be controlled.

Control of Hazards

A control is a method used to alleviate or lessen the harmful effects of the hazard. There are three types of controls—engineering controls, administrative controls, and the use of personal protective equipment (PPE). Safety professionals can use one or a combination of these types of controls to protect workers. The first and best type of control to use is engineering controls, which take the hazard away at its source. Engineering controls may include:

- substitution of a less-hazardous chemical for the one currently in use
- introduction of safeguards and interlocks
- the enclosure an open system.

Administrative controls are the next line of defense and may include:

- written policy to limit personnel access to hazardous areas
- hazard awareness training to affected personnel
- the institution of work–rest cycles or work rotation to minimize exposure time.

PPE can be used as a control to protect the worker only as a last defense. PPE does not eliminate the hazard. Examples of commonly used PPE to protect against physical hazards may include:

- safety glasses
- hearing protection devices
- gloves—leather, neoprene, antivibration, and so on
- clothing—fire resistant, leather, cooling, insulating, and so on

- safety shoes
- hard hats
- lead aprons
- welding goggles (further examples of PPE are discussed in consequent chapters).

All protective equipment are designed in accordance with specific standards—Occupational Safety and Health Administration (OSHA), American National Standards Institute (ANSI), National Fire Protection Association (NFPA), National Institute for Occupational Safety and Health (NIOSH), National Electrical Code (NEC), and so on. Every piece of equipment is marked with the certification number, regulation, or testing requirements. For example, approved safety goggles will have the inscription Z87.1-1968 on the frames. This means the goggle design was tested by ANSI, specification number Z87.1, issued in 1968.

Hazard control can be accomplished by following a step-by-step approach:

1. Recognize and identify the hazards, both the primary hazards and the initiating or contributing hazards.
2. Complete a risk assessment, which involves evaluating the severity of the hazard and the probability that it will or can occur.
3. Control the hazard through:
   - engineering controls (substitution, mitigation, and introduction of safeguards and interlocks)
   - the implementation of administrative procedures (training on processes and hazards)
   - the use of PPE (safety glasses, hearing protection devices)—the last line of defense.

Example: There is a woodworker in a carpentry shop making pine crates for shipping supplies. He is using a circular saw to cut wood for his project.
Question: Identify the primary hazard, the contributing hazards, and the controls that might be in use.
Answer:
- Primary hazard—sharp saw blade
- Contributing hazards—noise, sawdust, flying objects, kickback by the wood
- Controls—machine guards, training, safe work practices, safety glasses, hearing protection, leather gloves

---

Safety through Design

Over time, the level of safety achieved will relate directly to the caliber of the initial design of facilities, hardware, equipment, tooling, operations layout, work environment, and work methods and their redesign as continuous improvement is sought. The goal of continuous redesign and improvement of operations is to reduce the number of errors until operations are as error proof as human effort can make them. Efforts are made to reduce the number of safety decisions on the part of the employee; engineer out the hazard instead of having the employee take an action or use PPE for that hazard! The design stage offers the greatest opportunity to anticipate, analyze, eliminate, or control hazards. A good deal of information is gathered from lessons learned and machinery or plant operation history.

In the design and redesign process, management seeks to avoid, reduce, or eliminate the probability and severity of a hazard being realized and causing an incident. Companies should apply the following priorities to design and redesign processes (in priority order):

1. designing for minimum risk—ideal
2. incorporation of safety devices
3. providing warning devices
4. developing operating procedures and employee training programs
5. use of PPE—last resort

The safety professional can influence the design of the workplace and work methods at three critical points: in the preoperational, operation, and post-incident stages. Use of post-incident lessons learned and mishap investigations from other, similar processes can help justify correct initial design or redesign.

Example: As a safety professional, you have just heard about a very serious mishap or accident at an industrial facility similar to yours. The flywheel of a large stamping machine had a safety cage over the wheel itself but did not extend far enough, and some of the moving belt was exposed. The workers had no reason to be near that side of the machine, so it was considered of low risk by the machine manufacturer. Inevitably, one worker who dropped a tool went over to that side of the machine and bent down to retrieve it. The neckerchief around his shirt collar became caught in the belt and he was strangled.
Question: What actions should you as the safety professional take?
Answer:
- Cordon off the area, assign a temporary guard, post a sign, and instruct workers on the hazard.

- Contact OSHA and the machine manufacturer to see if there are records of this happening with this machine elsewhere.
- Investigate and inspect any similar machines and work procedures (neckerchiefs, training, hot environments, etc.).
- Contact the machine manufacturer for a retrofitted machine guard for the hazard.

Safety through design is defined as the integration of hazard analysis and risk assessment methods early in the design and engineering stages and the taking of the actions necessary so that the risks of injury or damage are at an acceptable level.

- hazard analysis—starts with a thorough job analysis to understand the process, procedure, individual steps, and the events of a job or task. Undesirable events are identified during this analysis, and hazards are defined and then analyzed or assessed as to their risk for damage or injury.
- risk assessment—once a hazard is identified and understood, the risk is assessed for the probability that it may occur and the severity if it does occur. Risk assessment must address each contributing hazard as well as the primary hazard.

Companies should take a proactive stance regarding safety through design. In implementing a safety-through-design process, companies should establish clear-cut objectives, assess hazard probability and severity, conduct hazard analysis and risk assessment, establish design review procedures, and use project checklists. Any time there is an incident, the most important step a safety professional must perform is to document, document, document! They should also examine work practices and systems more closely as causal factors rather than assuming worker behavior is the cause of an incident.

Job Hazard Analysis

Job analysis is done as part of a job hazard analysis (JHA). The safety professional must understand the process, the material used, how the equipment works, and how the worker interfaces with the equipment. Workers participate in the analysis by explaining how, when, where, what, and how often they conduct a step in the process; what they perceive as the hazards; and other details. The safety professional must work with the first-line supervisor and:

- observe the worker during the process or observe the machinery
- break each job down into the smallest tasks

- record each task and note the potential for any undesirable events
- determine what hazards exist and recommend controls.

A safety professional cannot just start at one end and work through the plant evaluating the jobs. There has to be a priority system for the evaluations, with the most hazardous areas assessed first. Prioritized candidates for a job analysis, in order of importance, are those jobs that:

1. involved a fatality—can be reevaluated as part of the investigation
2. resulted in injuries or illnesses
3. caused lost work time or compensation claims
4. have had repeated small accidents
5. have had near-misses
6. are new operations.

Physical hazards commonly identified during a job analysis include:

1. caught in, under, or between
2. fall to same level
3. fall to different level
4. struck by flying object
5. struck by falling object
6. strike against a stationary object
7. strains, sprains, or pull from pushing, pulling, bending, lifting, or twisting
8. cuts, lacerations, or contusions
9. struck by moving object
10. high noise levels
11. contact with energized equipment
12. ionizing radiation (X or gamma rays, alpha or beta particles)
13. non-ionizing radiation (illumination, infrared, ultraviolet, lasers)
14. environmental extremes (hot or cold)
15. ergonomic stresses
16. fire and explosion hazards.

Chemical or biological hazards identified during a job analysis include:

1. overexposure to dust, fumes, mists, vapors, or gases
2. skin irritation from oils, solvents, or detergents
3. molds, bacteria, or viruses.

An example of results obtained from performance of a job hazard analysis is provided in textbox 1.1.

Textbox 1.1. Example Job Hazard Analysis Form

Job Location: **Analyst:** **Date:**
Metal Shop   Joe Safety   April 21, 2011

Task Description: Worker reaches into metal box to the right of the machine, grasps a 15-pound casting, and carries it to grinding wheel. Worker grinds 20 to 30 castings per hour.

Hazard Description: Picking up a casting, the employee could drop it onto his foot. The casting's weight and height could seriously injure the worker's foot or toes.

Hazard Controls:
1. Remove castings from the box and place them on a table next to the grinder.
2. Wear steel-toe shoes with arch protection.
3. Use protective gloves that allow a better grip.
4. Use a device to pick up castings.

Summary

Physical hazards pose a very real and dangerous problem at the worksite. It is important for safety professionals to anticipate, recognize, evaluate, and remediate these types of hazards through using engineering controls and administrative controls and designating appropriate PPE.

Review Questions

1. Define a hazard.
2. What are three categories of physical hazards?
3. What are the different types of controls? Give an example of each.
4. Explain the principle of safety through design.
5. The safety professional can influence the design of the workplace and work methods at three critical stages. Name them.
6. List the steps involved in performing a JHA.

2

Layout, Construction, and Maintenance of Facilities

Building and Facility Layout

THE SAFETY-THROUGH-DESIGN PRINCIPLE discussed in chapter 1 can be carried out in the construction and renovation of the building and facility layout. Numerous accidents, occupational diseases, explosions, and fires can be prevented by carefully planning the design, location, and layout of a new facility or alterations to an existing one.

It is ideal for health and safety professionals to conduct a safety and health study of proposed construction in the developmental stages to remove hazards and reduce risks. Bear in mind the role of workers, the function of machines, and the flow of materials.

Four major factors should be considered in facility design and layout:

1. general design of the workplace
2. compliance with appropriate codes and standards
3. size, shape, and type of buildings, processes, and personnel facilities needed
4. safety procedures and fire protection standards required.

When planning outside facilities, safety and health professionals should ensure that issues of worker safety are incorporated into designs for company grounds; shipping and receiving facilities; and all roadways, walkways, trestles, and parking lots.

Facility layout of buildings and facilities should permit the most efficient use of materials, processes, and methods and minimize the hazards of fire and explosions. Layout of equipment can be done with the use of detailed flow sheets and must be designed to ensure maximum efficiency and safety for workers.

Both daylight and electric lighting can supply a facility's lighting needs. Proper illumination can help to reduce accidents, minimize hazardous areas, and make buildings and grounds more secure. Industrial designers and managers are paying more attention to the interactions of color, lighting, and human behavior. The light-reflectance value (LRV) of color refers to its effect on light, which can contribute to workers' abilities to see a task or identify color-coded materials. Colors also reflect regional and gender preferences, affect employees' morale, alter workers' perceptions of their surroundings, and provide an effective way to make hazardous items and safety signs more noticeable.

Examples of ANSI/OSHA color codes for marking physical hazards and identification of certain equipment include:

red—fire hazards, fire protection (hydrants, fire-exit signs, hoe locations), danger, emergency stops on equipment, danger safety cans or other portable containers of flammables, lights at barricades

orange—dangerous parts of machinery, such as energized equipment, exposed cutting devices, insides of guards

yellow—caution; hazards from slipping, falling, and so on; hazards from material handling equipment, such as cranes and lift trucks; radiation hazards (along with magenta and black)

green—first aid and safety equipment locations (e.g., emergency eye wash units)

blue—information that is not safety related; warning or caution limited to warning against starting, using, or moving equipment under repair

magenta (reddish-purple)—radiation protection

black and white—housekeeping and traffic aisles, stairways, and directional signs.

ANSI color-coded signs include:

danger—red oval in top panel, black or red lettering on white background in lower panel

caution—yellow background color with black lettering

general safety—green background on upper panel, black or green lettering on white background on lower panel

fire and emergency—white letters on red background
information—blue letters on white background.

Walking and Working Surfaces

Walking and working surfaces include floors, stairs, ladders, scaffolding, ramps, aisleways, platforms, and all other surfaces that are used by workers in the performance of their jobs. When addressing walking and working surfaces, OSHA also includes exit and egress requirements and floor and wall openings.

There are specific design requirements for these installations. General rules for walkways and working surfaces include:

- Surfaces must be nonslip or slip resistant and free from standing water.
- Railings must be provided if the level is four feet or more above the lower level.
- Areas must not be obstructed and remain trip hazard free and clutter free.
- The surfaces should provide good footing (concrete, asphalt, tile, etc.).
- Ice and snow removal must be considered for outdoor walkways.
- There must be sufficient width to move through aisles (three feet wider than the widest load).

Exits

The number and location of exits in a building is governed by the *Life Safety Code* issued by the National Fire Protection Association (NFPA, 2009b).

- The number of exits may be determined by the hazards in the workplace. In general, in high-hazard areas, no part of a building or floor should be more than seventy-five feet from an exit.
- All exit doors must open in the direction of intended travel.
- Exits must have appropriate signs and be illuminated.
- Emergency exits must never be blocked or locked from the inside.

Openings and Holes

OSHA standards differentiate between openings and holes. A hole is greater than one inch but less than twelve inches in its smallest dimension. An opening, then, is more than twelve inches in its smallest dimension. Tools and equipment can fall through holes, but people can fall through floor openings. All must be guarded. Temporary openings pose the greatest hazard. Openings where there is a drop of four feet or more must be guarded.

Construction of Facilities

Use prequalification criteria to aid in the selection of a contractor who is both qualified and has the safety and health programs to meet client demands. Make sure to verify how many OSHA violations or citations have been levied against the bidding contractors. Safety considerations should be calculated into the costs of construction from the beginning of the project. Ensure that job analysis (JA) or job hazard analysis (JHA) is included in project specifications for high-hazard operations where the probability of equipment failure or personnel injury or death is significant. Accident histories and lessons learned are useful in determining the potential for hazards.

To ensure safety on facility construction sites, companies should *not* rely on building codes and construction safety standards alone. (Everything must be spelled out in the contract!) Contractors are responsible for training their workers in safe working habits, good housekeeping procedures, worksite safety, and use of personal protective equipment (PPE). All construction sites must be restricted to authorized personnel only. The company should ensure that the construction contract contains at least minimum safety, health, and equipment requirements on the contractor's part and that it provides for an effective safety program. The safety professional can analyze the project and estimate the accident frequency potential in order to devise an appropriate accident prevention program.

The construction site itself must be made as safe as possible to protect workers and equipment from various hazards. All heavy machinery, trucks, other mobile equipment, hoisting apparatus, and conveyors must have safety devices and be operated only by trained personnel. Workers must be trained in the safe use of flammable liquids, operation of power machinery and tools, and proper methods for erecting steel. Employees must wear proper protective equipment and observe general health and safety practices. Again, ensure all of this is written in the contract!

In excavation work, management must do a careful survey of the site to evaluate potential hazards involving underground utilities, soil conditions, and surrounding structures. Hire a special contractor for excavation work done within or adjacent to a building or done lower than wall or column footings and machinery or equipment foundations.

Special Considerations for Construction

Construction of ladders and scaffolds should conform to the provisions of the appropriate state or provincial or local codes. Workers should have a safe access to the scaffold, and the structure should be inspected and tested regularly.

Ladders

Portable ladders are constructed in accordance with ANSI standards. All portable ladders should be inspected routinely and taken out of service if damaged. Ladders are constructed of wood, fiberglass, or aluminum. Metal ladders should never be used in electrical work.

Fixed ladders over twenty feet in length must be provided with cages built to OSHA standards, with the exception of ladders on towers, water tanks, and chimneys. With those installations, a rest platform is required every 250 feet, as well as ladder safety rails to clip on fall arresting devices.

Scaffolds

A scaffold is defined as a temporary, elevated working platform used to support workers and material during construction and/or maintenance activities. Scaffolding is defined as the wood or metal framework that supports the scaffold. Scaffolds more than ten feet above the ground must have railings and toe boards on all open sides and ends. Employees must wear fall protection while erecting scaffolds.

Wood scaffolds greater than sixty feet in height must be designed by a qualified engineer. Tubular metal scaffolds greater than 125 feet in height must be designed by a registered professional engineer. OSHA provides details for construction requirements and use of scaffolds.

Hoists

Material and personnel hoists can be erected inside the building or in outside towers. Never permit personnel to ride on a material hoist or permit work in or on the hoistway while the hoist is in operation. All hoists should have a good signal system and be guarded to prevent material from falling on workers.

Openings and Holes

Again, all floor and roof holes, skylights, and openings into which people can walk must be guarded with enclosed guards or covered with material and bracing strong enough to support any load. Stairways, ladderways, entrances, and wall openings should be guarded to prevent workers from walking into these openings.

Facility Maintenance

Sloppily maintained facilities can lead to accidents, occupational diseases, explosions, and fires. Good housekeeping and proper maintenance and upkeep are an easy control to avoid serious personal injury and property damage.

Facility maintenance includes:

- proper long-term care of the buildings, grounds, and equipment
- routine care to maintain service and appearance
- repair work required to restore or improve service and appearance.

The safety and health professional should be involved in maintenance plans and notify maintenance personnel of hazards or faulty equipment that need attention.

Example: During routine safety inspection rounds, the safety manager sees oil dripping from a machine, causing a slip hazard on the floor. It may also pose an environmental hazard if it gets into floor drains. **Question:** What actions should you as the safety professional take? **Answer:** Maintenance would be notified to check gaskets and seals to eliminate the leaking oil.

Maintenance inspectors should establish a regular inspection program of the facility and grounds and notify the safety manager of any safety hazards they find. Many hazards are discovered during routine maintenance inspections or routine maintenance tasks. Maintenance inspections normally include:

- walls and floors—inspected for damage, defects, and wear and regularly repaired or replaced
- roof-mounted structures—roof damage can quickly lead to structural damage of other parts of the building and equipment
- tanks and towers—maintained for fire protection and to prevent serious accidents due to structural failure
- stacks and chimneys—inspected at least once every six months
- platforms and loading docks, concrete sidewalks and drives, and driveways—inspected for damage or wear and repaired on a regular basis

- underground utilities—only trained, closely supervised personnel using proper protective equipment and other safety devices should do inspection and maintenance of underground utilities
- electrical fixtures—maintenance workers should replace faulty lamps, repair broken fixtures, and dispose of all lamp-related refuse in special containers (fluorescent bulbs may contain small amounts of mercury); workers are also responsible for "relamping" or replacing burned-out bulbs to maintain adequate lighting levels in the building
- stairways and exits—inspected for such conditions as inadequate design or construction, improper handrails, poor lighting or housekeeping, and faulty treads or damaged surfaces; keep exits clear and well lighted.

Safety hazards identified by the maintenance inspector or maintenance personnel should be brought to the attention of the safety manager. The safety manager may want to put a procedure in place to gather this safety hazard information (report form, e-mail template, etc.). A plan of action to correct these hazards can be generated, prioritizing the most critical hazards for timely correction or an interim fix.

Example: During the routine maintenance of an automated paper punch in a bindery, it is found that the plastic switch cover is missing and the wiring around the button is now exposed. The maintenance person has to order a new switch assembly because they have none in stock. This may take one to two weeks. The machine has to remain in service. They notify the safety manager, who is asked to recommend a work-around to safely use the switch. After going to the site, the safety manager recommends using a piece of sturdy cardboard to cover the switch, with a small opening to access the button. This is taped over the switch box. A sign is posted warning users of this temporary fix and for users to not remove the cardboard. The safety manager ensures that the switch has been ordered and sends a memo to management indicating the interim safety action taken.

Safety training is also required for maintenance staff:

- Management should select maintenance crews for their experience, alertness, and mechanical abilities and train them in accident prevention.
- Maintenance workers must dress appropriately for each job, use proper PPE and tools, and know all pertinent safety procedures and practices that apply to a firm's operations.

- To prevent accidents and injuries from grounds maintenance tools and machines, choose equipment for specific jobs and train workers in their particular responsibilities (line trimmers, lawn mowers, hedge clippers, etc.).
- Workers must know the safe-practice rules for handling and operating electric-powered hand-tools and gasoline-powered equipment.
- Workers must be carefully trained in the proper operation of utility tractors and snow throwers and must observe all safety precautions and manufacturer's instructions.
- Pesticides must be carefully selected, used, stored, and disposed of to prevent accidents and injuries. Personnel applying pesticides must be certified! Workers should protect themselves by using PPE and be trained in spill and decontamination procedures.

Computerized predictive maintenance (CPM) can reduce employees' exposure to hazards, decrease equipment downtime, make the most of maintenance expenditures, and create efficient schedules. The main benefits to CPM include:

- Maintenance diagnostic technology includes a number of methods to detect facility and grounds problems before they become serious threats to a company's operations. (The machine tells you when it is sick!)
- Supervisors can set up preventive maintenance and inspection plans for critical equipment and machinery that might seriously affect worker safety and health.
- Maintenance supervisors and crews can keep up to date on the developments in their trades.

Industrial Sanitation and Personnel Facilities

If the facility is large enough, management should consider establishing a separate department with its own director to ensure a sanitary work environment. Five industrial health areas must be kept sanitary and well equipped for employee health and convenience:

1. potable water supplies
2. sewage and garbage disposal
3. personal service facilities (drinking fountains, washrooms and locker rooms, showers, toilets, and janitorial service)
4. food service
5. heating, cooling, and ventilation.

General rules for water and sewage sanitation include:

- approved piping and storage system (tanks)
- good housekeeping, personal cleanliness, and a good inspection system
- drinking water that meets all regulatory standards of purity to avoid exposing workers to waterborne contaminants
- private water sources that are surveyed routinely to protect workers' health (remote locations may require wells)
- plumbing and fixtures that are disinfected before being put into service, either for the first time or after repairs
- filtration and chemical purification with disinfectants (these are the most practical methods for treating industrial private water supplies).

Organizations must adhere to state and local ordinances in disposing of and recycling sewage, waste, and garbage. Companies providing food services for their employees must:

- arrange for proper disposal of food refuse
- ensure that all garbage-collection areas and receptacles meet state and local codes
- employ pest control procedures to address who takes care of insect, rodent, and nuisance bird infestations.

All personal-service facilities contributing to employee health, such as drinking fountains, washrooms, locker rooms, showers, and toilets, should be kept clean and safe for workers' use. Companies must:

- maintain washrooms, showers, locker rooms, toilets, hand-washing facilities, and hand-drying mechanisms in a sanitary condition
- provide sinks to wash and dry PPE, clean storage for PPE, and storage for protective clothing; laundry facilities for contaminated clothing may also be necessary
- possibly collect and treat contaminated wastewater from cleaning or laundering contaminated clothing or PPE (e.g., lead paint removal workers).

Company health and safety staff can educate workers on proper nutrition and its relation to job performance and health. The five main types of industrial food service include cafeterias, canteens or lunchrooms, mobile canteens, box lunch services, and vending machines. Food-service staff must receive careful training in sanitation and food-handling methods to prevent food contamination, cross-contamination, and the transmission of foodborne illnesses.

Reasonable Accommodation for Workers with Disabilities

The Americans with Disabilities Act (ADA) of 1990 mandates that employers provide "reasonable accommodation" for workers with disabilities when feasible. Supervisors and other management personnel must be trained on the requirements for accommodating this group of workers.

Employers who deny disabled workers a job must prove that these workers would endanger themselves or others, cannot meet job requirements, or that reasonable accommodation of the workplace or job is not feasible. Job announcements must indicate physical requirements, if that is a specific condition of the job, such as being able to routinely lift fifty-pound boxes four feet off the ground (package delivery companies, for example, have such stipulations in their job qualifications). Various federal, state, and local laws also mandate affirmative action programs for the hiring and advancement of disabled persons. The law defines those types of disabled job seekers:

- disabled individual
- disabled veteran
- qualified disabled individual

This definition may impact the hiring priority and requirements for a company. Some companies have a policy for hiring disabled veterans, for example.

"Reasonable accommodation" as it pertains to the physical or mental limitations of an otherwise qualified disabled applicant or employee include modifying equipment or changing the job description. To ensure this is done thoroughly and appropriately:

- The safety and health professional can work with the equal employment opportunity (EEO) coordinator to evaluate reasonable accommodation and ensure compliance with government affirmative action regulations.
- Management must evaluate each disabled job applicant to place the right person in the right job so the impairment is not a factor of job performance.
- More stringent standards may need to be developed to ensure safe working conditions for these employees.
- Employers must evaluate jobs in terms of physical requirements, working conditions, health hazards, and injury hazards to eliminate or control job risks that might endanger disabled workers.

- Companies can also contact self-help and support groups, governmental agencies, and private agencies to learn more about how to accommodate a job for a disabled employee and how to work with disabled persons.

ADA addresses the following general areas concerning safety:

- The job would put the individual in a hazardous situation.
- Other employees would be placed in a hazardous situation if the person were on the job.
- The job requirements cannot be met by an individual with certain physical or mental limitations.
- Accommodation of the job cannot readily be accomplished.

The four factors employers must consider when evaluating a job for employees with one or more disabilities are:

1. physical requirements
2. working conditions, including emergency situations
3. health hazards
4. physical hazards.

The five important factors to consider when designing access to facilities for disabled workers are:

1. cafeteria, washroom, and restroom facilities
2. width of doors, ramps, elevators, and emergency exits
3. height of plumbing fixtures, desks, work stations, and tabletops
4. phones, emergency alarms, and emergency lighting
5. drinking fountains.

This information is also applied to returning an employee to work while recovering from a mishap or when coming off worker's compensation.

OSHA has a federal return-to-work program called SHARE (Safety, Health, and Return-to-Employment). It encourages companies to place workers into jobs that they can do, if not their original job, so they can return to employment and come off worker's compensation. This may require determining what job this worker is now physically capable of and, if a disability exists, where they can work safely.

Example: Mr. Jones, the plant electrician, slipped off a ladder while changing a lightbulb and shattered his ankle when he fell. Mr. Jones had surgery and physical therapy but cannot stand or walk for long periods of time. He has been on worker's compensation and off work now for six months. His doctor evaluated him and said Mr. Jones cannot fulfill his job obligations as an electrician in the future because of his physical limitations. The doctor said Mr. Jones would be able to do a desk or administrative-type job with no problems. Under the plant's return-to-employment program, the next administrative job that comes open is reviewed for possibly placing Mr. Jones in that job. The safety manager is asked to review his case file and the job requirements to ensure Mr. Jones can do this work safely within the limits of his physical condition.

Summary

The overall physical safety and health of workers can be accomplished if great care and concern are put into the physical layout, upkeep, and design of buildings. A proper building may provide workers with ergonomically designed work stations, ease of access to the workspace, and open clean areas. Many safety and health hazards can be eliminated or reduced through careful planning.

Review Questions

1. List the four major factors that should be considered in facility design and layout.
2. Why is proper lighting so important?
3. What are some of the responsibilities of the contractor?
4. Why should the safety professional coordinate with the facility maintenance and repair personnel?
5. Explain ADA. What are the employer's responsibilities?
6. What is SHARE?

3

Safeguarding, Lockouts, and Tagouts

During the period 1992–99, just two categories of events were jointly responsible for nearly three-quarters of all fatal workplace amputations. Workers being caught in or compressed by equipment accounted for 42 percent of the 171 fatal amputations, most of these (62 cases) were associated with workers being caught in running equipment or machinery. Similar to fatalities, two-thirds of all nonfatal workplace amputations reported in 1999 resulted from workers being caught in running equipment or machinery.—Bureau of Labor Statistics, 2003

Safeguarding

THE BASIC PURPOSE OF SAFEGUARDING of machines and equipment is to prevent contact of the human body with dangerous parts of machines. It's just another way design aids in the control or elimination of physical hazards. Methods of machine guarding vary greatly, depending on the type of machine, function of the machine, and the hazard to the operator.

Unguarded machines present two different types of hazards:

1. mechanical—hazards arising from the motion or operation of the machine
2. nonmechanical—hazards unrelated to machine motion, such as electrical hazards.

Figure 3.1.   Bench grinder missing machine guards.

Mechanical Hazards

Machine Motions

1. Hazards can be caused by rotating machinery in two ways:
   a. Protruding objects on rotating machinery, such as burrs, screws, bolts, rough surfaces, or the rotating motion itself tend to "grab" hair and clothing, causing the object to be pulled into the machine.
   b. In-running nip points, such as gears, chains, pressure rollers, and sprockets, when machine parts rotate toward each other or toward a stationary object, can cause a severe hazard by pulling objects into rotating parts or crushing the object against a stationary component.
2. Hazards caused by reciprocating motions follow a back-and-forth or up-and-down pattern. A worker can be struck by the reciprocating component or caught between the moving component and a stationary object. An example would be an automated woodworking machine equipped with a moving table. The worker could be struck by the moving table or caught between the moving table and the wall.

3. Hazards from transverse motions follow a continuous, straight line, such as a conveyor belt. The worker may be struck or caught in a nip or shear point and can be carried away or dragged by moving conveyor belts or similar systems.

Machine Actions

The specific process performed by the machine can present hazardous conditions:

- Cutting action is one that cuts or removes material from an object, such as saws, grinding wheels, planers, or drills.
- Punching, shearing, and bending action occurs as a powered ram or slide contacts metal, wood, or other material for the purpose of blanking, drawing, stamping, or trimming. Examples are power presses, shears, and benders.

Nonmechanical Hazards

- Electrical power sources can present a hazard if unguarded. Contact with improperly grounded sources or damaged wires can cause severe injury or death.
- High-pressure systems can malfunction and release pressurized gases, blow out gauges, rupture lines, and produce explosions or shrapnel hazards.
- Noise is an inherent hazard of virtually all machinery. Properly designed and installed guards can assist in reducing noise levels (dampers), but some guards can actually increase noise levels due to vibration or can channel the noise like a speaker.
- Chemical emissions can also be reduced by the use of guards. Lubricating oils, cutting fluids, cleaning fluids, and other process chemicals can easily be misted into the air by the motion of the machinery. Barrier guards or shielding can protect workers from this exposure.
- Flying objects, such as pieces of wood from cutting, metal trimmings from shearing, and objects that may become entrained in rotating machinery can present severe hazards. Barriers are usually used to guard against flying objects.

Common Safeguarding Methods

- guards—fixed, interlocked, adjustable, and self-adjusting; usually physical impediments

- devices—sensing devices, pullback mechanisms, restraints, safety controls, and gates; usually part of the machine mechanism itself
- automatic and semiautomatic feeding and ejecting methods—limits the involvement of the operator
- location and distance—moving the hazard out of reach of the operator; place the operator in a control room
- safety accessories—push-blocks for saws, nonkickback devices, jigs, special tools to keep the operator's hands out or away from the machine.

OSHA 29 CFR 1910.212 (2010) states that "guarding shall be provided to protect the operator and other employees in the machine area from hazards such as those created by point of operation, ingoing nip points, rotating parts, flying chips, and sparks." Design of guards is usually the responsibility of the manufacturer, but guards can be lost or require replacement or redesign. Older machinery that did not come with a guard or machinery that is reconditioned may need to be reevaluated for guarding requirements.

General Requirements for Guard Design

- affixed to the machine and securely attached
- prevents contact between body parts and the machine hazard
- tamper-resistant design to prevent removal or disengagement
- durable construction
- protects workers from flying objects
- does not create new hazards
- does not create interference with the process
- allows for lubrication and maintenance
- allows for ventilation.

Periodic inspection of guards should be performed and maintenance personnel trained to replace guards properly if removed for repairs.

Example: Fox News (Associated Press, 2010) article reads,

N.Y. Janitor Crushed to Death in Trash Compactor
A maintenance man at a prominent western New York building was crushed to death after falling into a trash compactor, but no one realized it for weeks as his family and authorities searched for him, said police Sunday. Surveillance video shows John Adams tumbled into a compactor trying to retrieve a fallen trash bin at the One Niagara building, and the mechanism "prevented him from being able to escape and made it

impossible for co-workers or others to realize what had occurred," Niagara Falls, N.Y., police said in a release.

Question: What questions should a safety professional ask about the machinery?
Answer:
- Were there any signs posted around the machinery warning employees about entrapment?
- Were there any fail-safes, sensors, or alarms inherent in the trash compactor? If so, why didn't they go off?
- Why weren't there any barriers or guards stopping a person from falling into the equipment?
- What steps need to be taken to prevent a similar accident in the future?

Lockout and Tagout (LOTO)

OSHA (2007a) defines LOTO as

> specific practices and procedures to safeguard employees from the unexpected ... startup of machinery and equipment, or the release of hazardous energy during service or maintenance activities.
>
> Approximately 3 million workers service equipment and face the greatest risk of injury if lockout/tagout is not properly implemented. Compliance with the lockout/tagout standard (29 CFR 1910.147) prevents an estimated 120 fatalities and 50,000 injuries each year. Workers injured on the job from exposure to hazardous energy lose an average of 24 workdays for recuperation. In a study conducted by the United Auto Workers (UAW), 20% of the fatalities (83 of 414) that occurred among their members between 1973 and 1995 were attributed to inadequate hazardous energy control procedures specifically, lockout/tagout procedures.

During maintenance and servicing of equipment or machines, safeguards may need to be removed or disengaged to provide access to the machine. During these operations, isolation and de-energizing of the equipment is required to prevent unexpected, injury-causing movement or startup of the machine. To isolate and de-energize to ensure the hazard is eliminated, we use LOTO procedures—physically *lock* the energy source with a locking device or place a caution *tag* warning workers to not touch the equipment.

Figure 3.2.   LOTO electrical box.

Figure 3.3.   Example of a LOTO work station.

Example: A 25-year-old male worker at a concrete pipe manufacturing facility died from injuries he received while cleaning a ribbon-type concrete mixer. The victim's daily tasks included cleaning out the concrete mixer at the end of the shift. The cleanout procedure was to shut off the power at the breaker panel (approximately thirty-five feet from the mixer), push the toggle switch by the mixer to make sure that the power was off, and then enter the mixer to clean it.

No one witnessed the event, but investigators concluded that the mixer operator had shut off the main breaker and then made a telephone call instead of following the normal procedure for checking the mixer before anyone entered it. The victim did not know that the operator had de-energized the mixer at the breaker. Thinking he was turning the mixer off, he activated the breaker switch and energized the mixer. The victim then entered the mixer and began cleaning without first pushing the toggle switch to make sure that the equipment was de-energized. The mixer operator returned from making his telephone call and pushed the toggle switch to check that the mixer was de-energized. The mixer started, and the operator heard the victim scream. He went immediately to the main breaker panel and shut off the mixer.

Within 30 minutes, the emergency medical service (EMS) transported the victim to a local hospital and then to a local trauma center. He died approximately four hours later (NIOSH, 1999).

Question: How could this fatality have been prevented?

OSHA's (2010) 29 CFR 1910.147 states that employers are required to develop, document, and utilize an energy control procedures program, to control potentially hazardous energy. The energy control program must specifically outline:

1. the scope, purpose, authorization, rules, and techniques to be utilized for the control of hazardous energy; and
2. the means to enforce compliance including, but not limited, to the following:
   - a specific statement of intended use of the procedure
   - specific procedural steps for shutting down, isolating, blocking, and securing machines and equipment to control hazardous energy
   - specific procedural steps for placement, removal, and transfer of lockout devices or tagout devices and other energy control measures
   - specific requirements for testing a machine or equipment to determine and verify the effectiveness of lockout devices, tagout devices, and other energy control measures
   - providing approved locking devices and tags

- training all workers who may be locking out equipment and are the *authorized* employees who conduct the program, and training affected workers who may come into contact with a lock or tag
- periodic inspections to ensure the program is enforced
- documentation.

Summary

The safety professional must be cognizant of the necessary and available controls for machinery, including safeguards and development and implementation of a well-devised LOTO program. It is imperative that workers are trained on the proper precautions to take when working with machinery of all kinds. Without these three very important safety controls, physical injury at the workplace is imminent.

Review Questions

1. What is the main purpose of safeguarding machinery?
2. Explain the two types of hazards unguarded machines present.
3. List the methods of safeguarding equipment.
4. Explain what is meant by LOTO.
5. What must an energy control program specifically outline?

4

Confined Spaces

OSHA (1996a) Hazards Information Bulletin
Asphyxiation Hazard in Pits: Potential Confined Space Problem
The Syracuse, New York Area Office brought to our attention the potential ex-
istence of asphyxiation hazards in pits that house the control valves for waterfall
and water fountain displays in shopping malls. The purpose of this bulletin is to
alert the reader that these pits may be permit-required confined spaces which are
regulated by 29 CFR 1910.146.

An employee entered a fountain pit through a 3 foot by 3 foot opening. He
descended 7 feet via a fixed ladder to the bottom of the pit to adjust the valves
which controlled the fountain's water flow. The employee was unable to exit the
space because he lost consciousness. A partner attempted to rescue him but was
also unable to exit the pit because of the onset of weakness. He was, however, able
to call security for assistance. A security guard and a passerby also attempted to
enter the pit, but quickly abandoned the rescue due to the rapid onset of dizziness.

The fire department was called in to perform the rescue operation. Both em-
ployees were rescued, treated, observed, and released. The oxygen levels in that
particular pit at the time the employees entered is unknown.

MANY WORKPLACES CONTAIN SPACES that are considered "confined" because their configurations hinder the activities of employees who must enter, work in, and exit them. A confined space has limited or restricted means for entry or exit, and it is not designed for continuous employee occupancy. Confined spaces include, but are not limited to, underground vaults, tanks, storage bins, manholes, pits, silos, process vessels, and pipelines (OSHA, 2007b).

OSHA uses the term "permit-required confined space" (permit space) to describe

> a confined space that has one or more of the following characteristics:
> 1. contains or has the potential to contain a hazardous atmosphere;
> 2. contains a material that has the potential for engulfing an entrant;
> 3. has an internal configuration such that an entrant could be trapped or asphyxiated by inwardly converging walls or by a floor which slopes downward and tapers to a smaller cross-section; or
> 4. contains any other recognized safety or health hazard [such as unguarded machinery, exposed live wires, or heat stress]. (OSHA, 2007b)

Confined spaces may be encountered in virtually any occupation, therefore their recognition is the first step in preventing fatalities. Deaths in confined spaces often occur because the atmosphere is oxygen deficient or toxic. Confined spaces should be tested prior to entry and continually monitored. More than 60 percent of confined-space fatalities occur among would-be rescuers, therefore a well-designed and properly executed rescue plan is a must (NIOSH, 1986). Many workplaces contain spaces that are considered confined because their configurations hinder the activities of any employees who must enter, work in, and exit them. For example, employees who work in process vessels generally must squeeze in and out through narrow openings and perform their tasks while cramped or contorted.

There are many instances where employees who work in confined spaces face increased risk of exposure to serious hazards. Confinement itself can pose an entrapment hazard. The confinement of the space may keep workers closer to such hazards as asphyxiating atmospheres or the moving parts of a mixer. OSHA uses the term *permit-required confined space* (permit space) to describe those spaces that both meet the definition of *confined space* and pose health or safety hazards.

Asphyxiation is the leading cause of death in confined spaces. The asphyxiations that have occurred in permit spaces have generally resulted from oxygen deficiency or exposure to toxic atmospheres. There have been cases where employees who were working in water towers and bulk material hoppers slipped or fell into narrow, tapering discharge pipes and died of asphyxiation due to compression of the torso. Also, employees working in silos have been asphyxiated as the result of engulfment in finely divided particulate matter (such as sawdust) that blocks the breathing passages. OSHA has, in addition, documented confined space incidents in which victims were burned, ground up by auger-type conveyors, or crushed or battered by rotating or moving parts inside mixers. Failure to properly follow lockout/tagout procedures or to de-energize equipment inside the space prior to employee entry was a factor in many of those accidents.

Employers and employees must be aware of the degree to which the conditions of permit-space work can compound the risks of exposure to atmospheric or other serious hazards. Further, the elements of confinement, limited access, and restricted airflow can result in hazardous conditions that would not arise in an open workplace. For example, vapors that might otherwise be released into the open air can generate a highly toxic or otherwise harmful atmosphere within a confined space. Unfortunately, in many cases, employees have died because employers improvised or followed "traditional methods" rather than following existing OSHA standards, recognized safe industry practice, or common sense.

The failure to take proper precautions for permit-space entry operations has resulted in fatalities as opposed to injuries. OSHA notes that, by their very nature and configuration, many permit spaces contain atmospheres that, unless adequate precautions are taken, are immediately dangerous to life and health (IDLH). For example, many confined spaces are poorly ventilated—a condition that is favorable to the creation of an oxygen-deficient atmosphere and to the accumulation of toxic gases. Furthermore, by definition, a confined space is not designed for continuous employee occupancy; hence little consideration has been given to the preservation of human life within the confined space when employees need to enter it.

It is the obligation of the employer to evaluate the workplace to determine if any spaces are permit-required confined spaces. It must first be determined whether a space is a confined space. If it is a confined space, then it must be determined if it is a permit-required confined space. If it is a permit-required confined space, then it must be determined whether full permit entry rules apply or less restrictive alternative entry rules apply.

In general, the Permit-Required Confined Spaces Standard requires the employer to evaluate the workplace to determine if any spaces are permit-required confined spaces. If permit spaces are present and workers are authorized to enter such spaces, a comprehensive permit-spaces program must be developed and implemented. This is an overall plan/policy for protecting employees from permit-space hazards and for regulating employee entry into permit spaces. The OSHA standard includes detailed specification of the elements of an acceptable permit-spaces program [29 CFR 1910.146(d); OSHA, 2007b]. Permit spaces must be identified by signs, and entry must be controlled and limited to authorized persons. An important element of the requirements is that entry be regulated by a written entry permit system and that entry permits be recorded and issued for each entry into a permit space. The standard specifies strict procedures for evaluation and atmospheric testing of a space before and during an entry by workers. The standard requires that entry be monitored by an attendant outside the space and that provisions be made for rescue in the event of an emergency. The standard specifies training requirements and specific duties for authorized entrants, attendants, and supervisors. Rescue service provisions are required, and where feasible, rescue must be facilitated by a non-entry retrieval system, such as a harness and cable attached to a mechanical hoist (OSHA, 2007b).

Remember that, when dealing with confined spaces, fatalities can occur as a result of encountering one or more of the following potential hazards:

- lack of natural ventilation
- oxygen-deficient atmosphere
- flammable/explosive atmosphere
- unexpected release of hazardous energy
- limited entry and exit
- dangerous concentrations of air contaminants
- physical barriers or limitations to movement
- instability of stored product (NIOSH, 1986).

In light of findings to date regarding occupational deaths in confined spaces, NIOSH (1986) recommends that managers, supervisors, and workers be made familiar with the following three steps:

1. recognition: Worker training is essential to the *recognition* of what constitutes a confined space and the hazards that may be encountered in them. This training should stress that death to the worker is the likely outcome if proper precautions are not taken before entry is made.
2. testing, evaluation, and monitoring: All confined spaces should be *tested* by a qualified person before entry to determine whether the confined-space atmosphere is safe for entry. Tests should be made for oxygen level, flammability, and known or suspected toxic substances. *Evaluation* of the confined space should consider the following:
   - methods for isolating the space by mechanical or electrical means (e.g., double block and bleed, lockout, etc.)
   - the institution of LOTO procedures
   - ventilation of the space
   - cleaning and/or purging
   - work procedures, including use of safety lines attached to the person working in the confined space and its use by a standby person if trouble develops
   - personal protective equipment (PPE) required (e.g., clothing, respirator, boots, etc.)

  • special tools required
  • communications system to be used.

The confined space should be continuously *monitored* to determine whether the atmosphere has changed due to the work being performed.

3. rescue: Establish *rescue* procedures before entry, and these must be specific for each type of confined space. Assign standby person for each entry where warranted. Equip the standby person with rescue equipment, including a safety line attached to the worker in the confined space, self-contained breathing apparatus, protective clothing, boots, and so on. The standby person should use this attached safety line to help rescue the worker. Practice the rescue procedures frequently enough to provide a level of proficiency that eliminates life-threatening rescue attempts and ensures an efficient and calm response to any emergency.

According to the U.S. Department of Energy (USDOE, 1994), the most efficient way to protect workers from workplace hazards is to first remove obvious hazards that can be eliminated without significant effort. As discussed throughout the book, hazards should be eliminated and controlled through the following hierarchy of methods:

  • engineering controls
  • administrative controls
  • PPE.

Often, physical hazards discovered through preliminary evaluations and site and facility walkthroughs can be eliminated without significant effort or cost. These hazards should be removed to the extent possible before actual work at the site begins. Examples of ways to eliminate physical hazards associated with the site include:

  • removing unnecessary debris
  • guarding exposed electrical wiring or sharp or protruding objects
  • securing objects near elevated surfaces and combustible materials
  • eliminating slippery surfaces, dangerous flooring, and uneven terrain.

Hazards that cannot be readily eliminated should be properly controlled through engineering and/or administrative means. The primary objective of these controls is to reduce worker exposure to safe levels, thereby avoiding the need for PPE.

Engineering Controls

Hazards subject to engineering controls generally include those that present a high potential for illness or injury to workers. These hazards present levels of concern in the following areas:

- frequency of hazard (i.e., how often such a hazard is likely to occur at the worksite)
- effect of hazard (i.e., whether exposure to such a hazard would result in an injury or illness)
- extent of injury or illness resulting from the hazard
- range of effect of the hazard.

Engineering controls, such as radiation shielding, are intended to address major hazards and are the preferred control method. These controls consist primarily of systems that are necessary to reduce worker exposure and prevent propagation of contaminants to "clean" areas. Other examples of engineered controls include process enclosures maintained at negative pressure with high-efficiency particulate air (HEPA)–filtered ventilation and surface-water drainage systems. Protection of the public though engineered controls should also take into consideration the safety and health of workers. For example, when designing or selecting systems for mitigating dispersal of contaminants to outside areas, attention should also be given to effects on workers within the contaminated zone. Area enclosures can concentrate airborne contaminants if not properly ventilated.

Administrative Controls

The purpose of administrative controls is to encourage safe work practices. This is first accomplished by controlling the movement of personnel within hazardous areas. Establishment and demarcation of exclusion areas and physical access controls will prevent workers from unnecessarily entering hazardous areas. These controls should also include operating procedures and training programs that address safety precautions to be followed by workers when working in hazardous areas. Ensure workers are certified for the particular equipment they are operating. Some standards prohibit the use of administrative controls as a means for controlling a hazard.

PPE

PPE is a common method used in hazard control. Selection of the most appropriate level of protection and combinations of respiratory protection and protective clothing will depend on:

- level of knowledge of onsite chemical and radiological hazards
- properties, such as toxicity, radioactivity, route of exposure, and matrix of the contaminants known or suspected of being present
- type and measured concentrations of the contaminants that are known or suspected of being present
- potential for exposure to contaminants in air, liquids, soils, or by direct contact with hazardous materials
- physical hazards
- climatic conditions
- biological hazards.

Based on the evaluation of potential hazards that will vary with confined-space activities, PPE should be selected for specific tasks and work areas (e.g., exclusion zone, contamination reduction zone). The specific PPE required for each work area and/or task should be determined and listed by a qualified safety professional.

PPE is divided into two broad categories: respiratory protective equipment and personal protective clothing. Both of these categories are incorporated into the four levels of protection (levels A, B, C, and D) based on the potential severity of the hazard. The following sections provide detail and explanation of those categories.

Levels of PPE

The specific levels of PPE and necessary components for each level have been divided into four categories according to the degree of protection afforded. General guidelines for use are:

level A: worn when the highest level of respiratory, skin, and eye protection is needed

level B: worn when the highest level of respiratory protection is needed but a lesser level of skin protection is needed

level C: worn when the criteria for using air-purifying respirators are met and a lesser level of skin protection is needed

level D: refers to work conducted without respiratory protection. This level should be used only when the atmosphere contains no known or suspected airborne chemical or radiological contaminants and oxygen concentrations are between 19.5 percent and 23 percent.

Level A PPE

Respiratory Protection

Level A respiratory protection is positive-pressure, full face–piece self-contained breathing apparatus (SCBA), or positive-pressure supplied air respirator (with escape bottle for IDLH or potential IDLH atmosphere).

Protective Clothing

Protective clothing provides maximum skin protection. It is used when the potential exists for splash or immersion by chemicals and/or radiologically contaminated liquids or for exposure to vapors, fumes, gases, or particulates that are harmful to skin or capable of being absorbed through the skin. This class of protection is acceptable for radiological work activities categorized as "high" involving pressurized or large-volume liquids or closed-system breach.

Level A protective clothing includes:

• totally encapsulating nonpermeable, chemical-resistant suit
• coveralls inner suit
• modest clothing under coveralls (e.g., shorts and T-shirt/long underwear)
• disposable gloves and boot covers (worn over fully encapsulating suit)
• boots, chemical resistant, steel toe, and shank (depending on suit construction, worn over or under suit boot)
• hard hat (under suit)
• hearing protection (as needed).

Other PPE

Other protective apparatus that may be used includes:

• cooling unit/system
• two-way radio communications
• cold-weather gear/clothing
• protection from biological hazards/pests.

Level B PPE

Respiratory Protection

Level B respiratory protection is positive-pressure, full face–piece SCBA or a positive-pressure supplied air respirator (with escape bottle for IDLH or potential IDLH atmosphere).

Protective Clothing

Level B protective clothing provides a high level of skin protection. It is used when the potential exists for contact with chemicals and/or radiologically contaminated liquids that could saturate and/or penetrate cloth coveralls (e.g., immersion or inundation of contaminants), but vapors, fumes, gases, or dust containing levels of chemicals harmful to skin or capable of being absorbed through the skin are not expected. This class of protection is acceptable for radiological work activities categorized as "high" involving pressurized or large-volume liquids or closed-system breach. Level B protective clothing includes:

- hooded one-piece, nonpermeable, chemical-resistant outer suit
- coveralls inner suit(s)
- modest clothing under coveralls (e.g., shorts and T-shirt/long underwear)
- outer chemical-resistant work gloves (rated for contaminants) taped to outer suit
- inner gloves of lightweight PVC or latex rubber taped to inner suit (cotton liners optional)
- chemical-resistant steel-toe boots taped to inner suit
- disposable outer boot covers (booties) taped to outer suit
- hard hat (as needed)
- hearing protection (as needed).

Other PPE

Other protective apparatus that may be used includes:

- cooling unit/system
- cold-weather gear/clothing
- protection from biological hazards/pests.

Level C PPE

Respiratory Protection

Level C respiratory protection includes an air-purifying respirator, full face or half mask, cartridge or canister equipped (MSHA/NIOSH approved).

Protective Clothing

Level C protective clothing provides a moderate level of skin protection. It is used when the potential exists for contact with chemicals and/or radiologically contaminated materials but when protection from liquids (chemical and/or radioactive) is not required. It is used when potential vapors, fumes, gases, or dust are not suspected of containing levels of chemicals harmful to skin or capable of being absorbed through the skin. This class of protective clothing is appropriate for most routine radiological work activities.

Level C protective clothing includes:

- coveralls
- modest clothing under coveralls (e.g., shorts and T-shirt/long underwear)
- rubber/chemical-resistant outer gloves rated for contaminant
- inner gloves of lightweight PVC or latex rubber
- safety glasses or safety goggles (not required with full-face respirator)
- face shield if splash hazard exists (not required with full-face respirator)
- steel-toe rubber boots
- outer disposable booties
- hood may be required for radiological work
- hard hat (as needed)
- hearing protection (as needed).

Other Level C PPE

Other Level C protective apparatus that may be used includes:

- cooling unit/system
- cold-weather gear/clothing
- protection from biological hazards/pests.

Level D PPE

Respiratory Protection

There is no Level D PPE required for respiratory protection due to the nature of the hazard.

Protective Clothing

Level D protective clothing provides a low level of skin protection. It is used when there is no potential for contact with hazardous levels of chemicals or radiological contamination. This level should not be worn in the exclusion zone or the contamination reduction zone. Oversight personnel not in zoned areas, as well as site visitors, may be required to wear Level D modified PPE.

Level D protective clothing includes:

- coveralls
- modest clothing under coveralls
- work gloves where appropriate
- PVC or latex rubber surgical/lightweight gloves when sampling or handling any potentially contaminated surface or item
- safety glasses or safety goggles
- steel-toe rubber boots where wet decontamination methods are required or steel-toe leather boots and outer boot covers
- hard hat.

Other Level D PPE

Other Level D protective apparatus that may be used includes:

- cold-weather gear/clothing
- protection from biological hazards/pests
- hearing protection.

Use of PPE

Written site-operating procedures for the use of PPE should include:

- training
- establishing work mission duration
- personal use factors

- fit testing
- donning and doffing
- in-use monitoring of personnel/equipment
- inspection before, during, and after use
- storage and maintenance
- upgrading/downgrading of PPE
- decontamination and disposal.

Summary

The dangers of confined space entry define the level of PPE that must be used. It is the safety professional's duty to assess the risks associated with a space prior to employee occupancy. Once the hazards have been assessed, controls can be put into place to prevent deadly situations.

Review Questions

1. What is a confined space?
2. What are specific hazards that can be associated with confined spaces?
3. Controls are necessary to prevent injury and death in confined spaces. Give several examples of such controls.
4. Define permit-required confined space.
5. List the four levels of PPE.

5

Noise

What Is Noise?

NOISE, OR UNWANTED SOUND, is one of the most pervasive occupational health problems. It is a byproduct of many industrial processes and results in adult-onset hearing loss, which has been described as the "fifteenth most serious health problem" in the world. Estimates of the number of people affected worldwide by hearing loss increased from 120 million in 1995 to 250 million in 2004. A majority of noise-induced hearing loss (NIHL) is caused by occupational noise exposure. It is estimated that, in the United States alone, about 9 million workers are exposed to time-weighted average (TWA) sound levels of at least 85 dB (decibels), and about 10 million have NIHL greater than 25 dB (Nelson, Nelson, Concha-Barrientos, and Fingerhut, 2005). High levels of noise can be a persistent occupational hazard with many adverse effects, including elevated blood pressure, reduced performance, sleeping difficulties, annoyance and stress, tinnitus, NIHL, and temporary threshold shift. The most serious of these adverse health effects is permanent NIHL resulting from irreversible damage to the inner ear. The inner ear can be injured by exposure to a brief but intense sound, such as an explosion, as well as from repeated exposure to excessive sound levels over time (HEI, 2008). Neitzel, Siexas, Ren, Camp, and Yost (n.d.) state, "Workers suffering from NIHL are denied the ability to converse normally with others, are limited in their ability to perceive audible warnings in the workplace, and often suffer other related health problems and decreased job performance."

Controls

Most hearing-loss-prevention professionals agree that the ideal protection of workers from hearing loss is through the implementation of engineering controls and administrative controls to limit exposure to workplace hazard. However, in many occupational settings, protecting the workforce from hearing loss ultimately depends upon the proper use of personal protective equipment (hearing protection devices) and the informed and voluntary actions of the workers (NIOSH, 1996).

Machine and design changes are the easiest way to implement engineering controls but also the most costly and usually not the most practical. Other engineering controls may include:

- machine isolation, enclosure, or use of barriers
- use of isolation, enclosure, or barrier to remove the worker from the noisy process
- replacement of noisy equipment with quieter equipment
- relocation of equipment
- vibration control, ensuring the equipment is properly mounted or balanced.

Administrative controls to lessen employee exposure to noise may include:

- implementation of a hearing conservation program
- training employees to know the possible health effects, hazard sources, and use of hearing protection devices
- standard operating procedures for noise-hazardous work operations
- labeling and signage of noise hazardous areas and equipment
- worker rotation
- limiting the amount of time spent exposed to noisy processes
- replacing old parts and equipment with less noisy counterparts.

When noise control measures are unsuccessful, infeasible, or until such time as they are installed, hearing protection devices are the only way to prevent hazardous levels of noise from damaging the inner ear. The responsibility to ensure that hearing protection devices are used correctly and consistently lies on the shoulders of the management team. The following is a list of commonly used hearing protection devices:

- expandable foam plugs
- premolded plugs

- reusable plugs
- canal caps
- earmuffs.

Hearing Conservation Program

OSHA's hearing conservation program is designed to protect workers with significant occupational noise exposures from hearing impairment even if they are subject to such noise exposures over their entire working lifetimes. The following information taken from OSHA (2002) summarizes the required component of OSHA's hearing conservation program for general industry. It covers monitoring, audiometric testing, hearing protectors, training, and recordkeeping requirements.

Required Monitoring

The hearing conservation program requires employers to monitor noise exposure levels in a way that accurately identifies employees exposed to noise at or above 85 dB averaged over eight working hours, or an eight-hour TWA. Employers must monitor all employees whose noise exposure is equivalent to or greater than a noise exposure received in eight hours where the noise level is constantly 85 dB. The exposure measurement must include all continuous, intermittent, and impulsive noise within an 80 dB to 130 dB range and must be taken during a typical work situation. This requirement is performance oriented because it allows employers to choose the monitoring method that best suits each individual situation.

Employers must repeat monitoring whenever changes in production, process, or controls increase noise exposure. These changes may mean that more employees need to be included in the program or that their hearing protectors may no longer provide adequate protection. Employees are entitled to observe monitoring procedures and must receive notification of the results of exposure monitoring. The method used to notify employees is left to the employer's discretion.

Employers must carefully check or calibrate instruments used for monitoring employee exposures to ensure that the measurements are accurate. Calibration procedures are unique to specific instruments. Employers should follow the manufacturer's instructions to determine when and how extensively to calibrate the instrument.

Audiometric Testing

Audiometric testing monitors an employee's hearing over time. It also provides an opportunity for employers to educate employees about their hearing and the need to protect it. The employer must establish and maintain an audiometric testing program. The important elements of the program include baseline audiograms, annual audiograms, training, and follow-up procedures. Employers must make audiometric testing available at no cost to all employees who are exposed to an action level of 85 dB or above, measured as an eight-hour TWA.

The audiometric testing program follow-up should indicate whether the employer's hearing conservation program is preventing hearing loss. A licensed or certified audiologist, otolaryngologist, or other physician must be responsible for the program. Both professionals and trained technicians may conduct audiometric testing. The professional in charge of the program does not have to be present when a qualified technician conducts tests. The professional's responsibilities include overseeing the program and the work of the technicians, reviewing problem audiograms, and determining whether referral is necessary.

The employee needs a referral for further testing when test results are questionable or when related medical problems are suspected. If additional testing is necessary or if the employer suspects a medical pathology of the ear that is caused or aggravated by wearing hearing protectors, the employer must refer the employee for a clinical audiological evaluation or otological exam, as appropriate. There are two types of audiograms required in the hearing conservation program: baseline and annual audiograms.

Baseline Audiogram

The baseline audiogram is the reference audiogram against which future audiograms are compared. Employers must provide baseline audiograms within 6 months of an employee's first exposure at or above an eight-hour TWA of 85 dB. An exception is allowed when the employer uses a mobile test van for audiograms. In these instances, baseline audiograms must be completed within 1 year after an employee's first exposure to workplace noise at or above a TWA of 85 dB. Employees, however, must be fitted with, issued, and required to wear hearing protectors whenever they are exposed to noise levels above a TWA of 85 dB for any period exceeding 6 months after their first exposure until the baseline audiogram is conducted.

Baseline audiograms taken before the hearing conservation program took effect in 1983 are acceptable if the professional supervisor determines that the audiogram is valid. Employees should not be exposed to workplace noise for 14 hours before the baseline test or wear hearing protectors during this time period.

Annual Audiogram

Employers must provide annual audiograms within 1 year of the baseline. It is important to test workers' hearing annually to identify deterioration in their hearing ability as early as possible. This enables employers to initiate protective follow-up measures before hearing loss progresses. Employers must compare annual audiograms to baseline audiograms to determine whether the audiogram is valid and whether the employee has lost hearing ability or experienced a standard threshold shift (STS). An STS is an average shift in either ear of 10 dB or more at 2,000, 3,000, and 4,000 hertz.

Additional Employer Responsibilities

The employer must fit or refit any employee showing an STS with adequate hearing protectors, show the employee how to use them, and require the employee to wear them. Employers must notify employees within 21 days after the determination that their audiometric test results show an STS. Some employees with an STS may need further testing if the professional determines that their test results are questionable or if they have an ear problem thought to be caused or aggravated by wearing hearing protectors. If the suspected medical problem is not thought to be related to wearing hearing protection, the employer must advise the employee to see a physician. If subsequent audiometric tests show that the STS identified on a previous audiogram is not persistent, employees whose exposure to noise is less than a TWA of 90 dB may stop wearing hearing protectors.

The employer may substitute an annual audiogram for the original baseline audiogram if the professional supervising the audiometric program determines that the employee's STS is persistent. The employer must retain the original baseline audiogram, however, for the length of the employee's employment. This substitution will ensure that the same shift is not repeatedly identified. The professional also may decide to revise the baseline audiogram if the employee's hearing improves. This will ensure that the baseline reflects actual hearing thresholds to the extent possible. Employers must conduct audiometric tests in a room meeting specific background levels and with calibrated audiometers that meet ANSI specifications of SC-1969.

Hearing Protection Requirements

Employers must provide hearing protectors to all workers exposed to eight-hour TWA noise levels of 85 dB or above. This requirement ensures that employees have access to protectors before they experience any hearing loss.
Employees must wear hearing protectors:

- for any period exceeding six months from the time they are first exposed to eight-hour TWA noise levels of 85 dB or above, until they receive their baseline audiograms if these tests are delayed due to mobile test van scheduling
- if they have incurred standard threshold shifts that demonstrate they are susceptible to noise
- if they are exposed to noise over the permissible exposure limit of 90 dB over an eight-hour TWA.

Employers must provide employees with a selection of at least one variety of hearing plug and one variety of hearing muff. Employees should decide, with the help of a person trained to fit hearing protectors, which size and type of protector is most suitable for the working environment. The protector selected should be comfortable to wear and offer sufficient protection to prevent hearing loss.

Hearing protectors must adequately reduce the noise level for each employee's work environment. Most employers use the Noise Reduction Rating (NRR) that represents the protector's ability to reduce noise under ideal laboratory conditions. The employer then adjusts the NRR to reflect noise reduction in the actual working environment.

The employer must reevaluate the suitability of the employee's hearing protector whenever a change in working conditions may make it inadequate. If workplace noise levels increase, employers must give employees more effective protectors. The protector must reduce employee exposures to at least 90 dB and to 85 dB when an STS already has occurred in the worker's hearing. Employers must show employees how to use and care for their protectors and supervise them on the job to ensure that they continue to wear them correctly.

Training

Employee training is very important. Workers who understand the reasons for the hearing conservation programs and the need to protect their hearing will be more motivated to wear their protectors and take audiometric tests. Employers must train employees exposed to TWAs of 85 dB and above at least annually in the effects of noise; the purpose, advantages, and disadvantages of various types of hearing protectors; the selection, fit, and care of protectors; and the purpose and procedures of audiometric testing. The training program may be structured in any format, with different portions conducted by different individuals and at different times, as long as the required topics are covered.

Recordkeeping

Employers must keep noise exposure measurement records for two years and maintain records of audiometric test results for the duration of the affected employee's employment. Audiometric test records must include the employee's name and job classification, date, examiner's name, date of the last acoustic or exhaustive calibration, measurements of the background sound pressure levels in audiometric test rooms, and the employee's most recent noise exposure measurement.

Employers also are required to record work-related hearing loss cases when an employee's hearing test shows a marked decrease in overall hearing. Employers will be able to make adjustments for hearing loss caused by aging, seek the advice of a physician or licensed healthcare professional to determine if the loss is work related, and perform additional hearing tests to verify the persistence of the hearing loss.

Summary

Noise continues to be one of the most prevalent occupational hazards worldwide. It is the safety professional's duty to provide the workplace with the proper tools to deal with occupational noise exposures. This could be in the form of development and enforcement of a hearing conservation program, making it policy to order new tools that are less noisy, enclosing noisy machinery, and ensuring workers are trained and provided with appropriate hearing protection devices.

Review Questions

1. Give an example of an engineering control for noisy machinery.
2. What type of administrative controls are effective in protecting worker hearing?
3. Why is training of workers so important?
4. List the different types of hearing protection devices.
5. What are the elements of a hearing conservation program?

6

Radiation

RADIATION SOURCES MAY BE FOUND in a wide range of occupational settings, including healthcare facilities, research institutions, nuclear reactor facilities, nuclear weapon production facilities, shipyards, construction, and various manufacturing settings. There are two main types of radiation: ionizing and non-ionizing.

Ionizing Radiation

The CDC (2006) defines ionizing radiation as any radiation capable of displacing electrons from atoms, thereby producing ions. OSHA 1910.1096(a)(1) states that ionizing radiation includes alpha rays, beta rays, gamma rays, X-rays, neutrons, high-speed electrons, high-speed protons, and other atomic particles. Ionizing radiation sources can pose a considerable health risk to affected workers (OSHA, radiation). The seriousness of health effects is dependent upon the amount of the time the worker is exposed (how long) and the dose of the exposure (how much). It can be extremely harmful, especially in high doses over a short period of time or in smaller doses over a long period of time. Radiation exposure may cause physical and biological damage to the cells, resulting in severe burns to the skin and eyes, DNA changes, or destruction in rapidly growing cells, such as blood cells, sex cells, and even an unborn child. Extensive exposure to radiation can lead to radiation sickness, cancer, and death.

Controls

There is no safe level of radiation exposure. It is the employer's responsibility to ensure that all employee exposures are kept below regulatory limits and that employees are not being exposed without necessity. Engineering controls that can be used to limit exposure to ionizing radiation sources include:

- installation of permanent shielding, barriers, and enclosures (medical X-ray rooms are constructed with lead-lined walls)
- substituting radioactive materials with a less-hazardous substance (design equipment to use less of the hazardous substance)
- installation of safety switches, locks, and interlocks (medical X-ray equipment will not engage unless a user is pressing the activation button).

The most important administrative control when dealing with ionizing radiation is to keep it as low as reasonably achievable (ALARA). It does not sound like much, but when dealing with radiation, it is imperative to keep all exposures to a minimum. This is accomplished by:

- limiting exposure time
- increasing distance from the source at every opportunity
- using remote handling devices when available
- making it policy and procedure to work quickly and efficiently (be prepared, have equipment ready)
- ensuring the proper use of labels and signs
- monitoring workers' radiation exposure (radiation dosimeters)
- proper training of radiation workers.

As in most instances, if the hazard cannot be abated, the use of PPE is recommended along with engineering and administrative controls. The main source of PPE when dealing with radiation is the use of personal shielding devices. Personal shielding devices come in the form of aprons, gloves, and blankets; even regular clothing can protect the wearer from certain types of radiation exposure. When dealing with alpha and beta types of radiation, safety glasses or goggles are also recommended. Always remember these three safety words when working around ionizing radiation—*time, distance,* and *shielding*!

Non-Ionizing Radiation

Non-ionizing radiation can be defined as radiation that has lower energy levels and longer wavelengths than ionizing radiation. It is not strong enough to affect the structure of atoms it contacts but is strong enough to heat tissue and can cause harmful biological effects (CDC, 2006). The most common injuries incurred by overexposure to non-ionizing radiation are eye injuries (burns, blindness) and skin burns. Common types of non-ionizing radiation are microwaves (MW), radio frequency (RF), visible light, infrared (IR), ultraviolet light (UV), and extremely low frequency (ELF). Common sources of ionizing radiation are presented in table 6.1.

Laser Hazards

Laser is an acronym that stands for "light amplification by stimulated emission of radiation." Lasers create an intense directional beam that concentrates energy (heat) in a small area, causing superheating of tissues. Overexposure to laser radiation can result in damage to the skin and eyes. Of the two, the eye is much more vulnerable to injury from lasers, since it is susceptible to retinal burns, cataracts, and even blindness (OSHA, 1991).

The various classes of lasers and the degree of hazard each poses are as follows:

class 1: Lasers are encased inside equipment; they post no hazard as long as the equipment is not disassembled (laser engravers, CD players).
class 2/2a: This class of laser is of a relatively low hazard. Lasers can pose a hazard if one stares into the beam (super market scanners).

Table 6.1.   Common Sources of Ionizing Radiation

Type of Non-Ionizing Radiation	Sources of Radiation
RF and MW	radios, cellular phones, the processing and cooking of foods, heat sealers, vinyl welders, high-frequency welders, induction heaters, flow solder machines, communications transmitters, radar transmitters, ion implant equipment, MW drying equipment, sputtering equipment, and glue curing
ELF	power lines, electrical wiring, and electrical equipment
Visible Light	lasers, electromagnetic radiation (too much or not enough)
IR	furnaces, heat lamps, and IR lasers
UV	sun, black lights, welding arcs, and UV lasers

class 3a: There is definite eye hazard if the beam is directed or focused into the eye (laser pointers).

class 3b: There is hazard if the beam is directed toward or reflected into view of the eye (research lasers).

class 4: There is eye hazard if the beam is directed, focused, or diffusely reflected into the eye. This class of laser also poses skin and fire hazards (medical, industrial, and research lasers).

Controls

Control measures must be implemented to "reduce the possibility of exposure of the eye and skin to hazardous laser radiation and to other hazards associated with the operation of lasers and laser systems. This applies during normal operation and maintenance by users, as well as by manufacturers during the manufacture, testing, alignment, servicing, etc. of lasers and laser systems" (OSHA, 1991).

There are three basic categories of controls useful in laser environments. These are engineering controls, administrative controls, and PPE. The controls are based upon the recommendations of the ANSI Z-136.1 standard. Engineering controls can consist of:

- enclosed beam paths
- blocking barriers
- remote controls
- beam attenuators
- interlocks
- panic buttons.

Administrative controls in a laser-rich environment must consist of:

- signs, labels, and postings warning employees of the type of laser and the necessary PPE
- access for authorized and essential personnel
- training on the hazards and controls for the type of laser being used
- development and enforcement of standard operating procedures
- designated laser control areas.

PPE is required when working with and near hazardous lasers. Ensure that personnel is trained on the use and provided with the necessary PPE to include:

- appropriately rated glasses or goggles
- gloves
- a lab jacket or coat.

As stated in 29 CFR 1926.102(b)(2), laser work and similar operations create intense concentrations of heat, UV, IR, and reflected light radiation. A laser beam of sufficient power can produce intensities greater than those experienced when looking directly at the sun. Unprotected laser exposure may result in eye injuries, including retinal burns, cataracts, and permanent blindness. When lasers produce invisible UV or other radiation, both employees and visitors should use appropriate eye protection at all times.

Determine the maximum power density, or intensity, lasers produce when workers are exposed to laser beams. Based on this knowledge, select lenses that protect against the maximum intensity. The selection of laser protection should depend upon the lasers in use and the operating conditions. Workers with exposure to laser beams must be furnished suitable laser protection. The basic requirements for protective eyewear as proposed in the ANSI Z-136.1 standard can be summarized as follows:

- Protective eyewear shall be worn whenever operational conditions may result in potential eye hazard.
- All laser protective eyewear shall be clearly labeled with the optical density value and wavelength for which protection is afforded.
- Protective eyewear should be comfortable, have adequate visibility (luminous transmission), and prevent hazardous peripheral radiation.
- Periodic inspection shall be made of protective eye wear to insure the maintenance of satisfactory filtration ability. This shall include inspection of the filter material for pitting, crazing, cracking, and so on and inspection of the goggle frame for mechanical integrity and light leaks (OSHA, 1991).

Radio Frequency and Microwave Radiation

OSHA (2005) states, "Radiofrequency and microwave radiation are electromagnetic radiation in the frequency ranges 3 kilohertz (kHz)–300 Megahertz (MHz), and 300 MHz–300 gigahertz (GHz), respectively. Research continues on possible biological effects of exposure to RF/MW radiation from radios, cellular phones, the processing and cooking of foods, heat sealers, vinyl welders, high frequency welders, induction heaters, flow solder machines, communications transmitters, radar transmitters, ion implant equipment, microwave drying equipment, sputtering equipment and glue curing."

According to NIOSH/OSHA (1979), the absorption of excessive amounts of RF energy may cause changes in the eye, the central nervous system, conditioned reflex behavior, heart rate, chemical composition of the blood, and the immunologic system. OSHA (2005) continues, "High power densities, RF/MW energy can cause thermal effects that can cause blindness, and sterility. Non-thermal effects, such as alteration of the human body's circadian rhythms, immune system and the nature of the electrical and chemical signals communicated through the cell membrane have been demonstrated. However, none of the research has conclusively proven that low-level RF/MW radiation causes adverse health effects."

NIOSH and OSHA (1979) recommends that precautionary measures be instituted to minimize the risk to workers from unwarranted exposure to RF energy. The following engineering controls, administrative controls, and PPE are recommended. Engineering controls might consist of:

- proper design and installation of shielding material
- providing automatic feeding devices, rotating tables, and remote materials handling to maximize distance between source and worker
- tuning the equipment electronically to minimize the stray power emitted.

Administrative controls to consider include:

- designation of restricted areas
- limiting amount of time personnel spend in RF areas
- RF safety training
- switching off equipment when it is not being used
- ensuring maintenance and adjustment of the equipment is performed
- posting warnings and information
- development and implementation of necessary medical surveillance programs tailored to the expected degree of employee use of RF equipment and potential for exposure to RF energy
- taking exposure measurements at regular intervals to ensure employee safety
- use of personal radiation monitors.

PPE is not usually required when dealing with RF radiation, however some companies may recommended or require certain types of personal shielding or barriers, which may be in the form of RF suits.

Summary

The easiest ways to limit exposure to radiation sources is through time, distance, and shielding. Limit the amount of time an employee spends in radiation areas, ensure employees are trained and know to increase their distance from the source as often as possible, and make sure they are using proper protective equipment. All of these steps follow the ALARA principle. It is the safety professional's duty to provide radiation training to employees and provide documentation of their exposures.

Review Questions

1. Define and give examples of ionizing radiation.
2. What health effects may occur from ionizing radiation overexposures?
3. Define and give examples of non-ionizing radiation.
4. What are the types of lasers? How do they differ?
5. What is meant by ALARA? How is it accomplished?

7

Ergonomics

What Is Ergonomics?

Aᴄᴄᴏʀᴅɪɴɢ ᴛᴏ ᴛʜᴇ CDC (2010), ergonomics is the scientific study of people at work. The goal of ergonomics is to reduce stress and eliminate injuries and disorders associated with the overuse of muscles, bad posture, and repeated tasks. This is accomplished by designing tasks, workspaces, controls, displays, tools, lighting, and equipment to fit the employee's physical capabilities and limitations. Fit the job to the employee, not the employee to the job!

Furthermore, OSHA refers to ergonomics as the interrelationship of workers and their work stations, also called human factors or human engineering. It is used by industrial engineers to promote production efficiency. It is a science that uses many disciplines to evaluate body stressors.

- anthropometry—considers the physical variables of employees (Mr. "Average Man" has changed over the years!)
- biomechanics—the moving body; looks at stresses put on body parts when lifting, pushing, pulling, twisting, reaching, and so on
- physiology—deals with body functions and processes to estimate metabolic rates and needs of the cardiovascular system for work tasks
- psychology—study of behavioral responses that include those influenced by the work environment.

Common musculoskeletal disorders include: chronic back pain, carpal tunnel syndrome, tendonitis, tenosynovitis, rotator cuff syndrome, sprains,

Reynaud's syndrome, and strains. Work-related musculoskeletal disorders (WMSDs) are caused by job activities and conditions, like lifting, repetitive motions, and work in confined areas. The chances of developing WMSDs increases when one of more of these physical tasks are performed:

- carrying heavy loads
- working on your knees
- twisting your hands or wrists
- stretching to work overhead
- gripping certain types of tools
- using vibrating tools or equipment
- doing activity that involves repetitive movement
- lifting heavy loads (NIOSH, 2007).

Work stations, assembly lines, work tasks, and machine operator stations are evaluated as part of the job analysis (JA) to identify hazards. Tasks that produce body stresses, awkward movements, or extremes of movement or require lifting or bending should be documented and reviewed. Each operation should be investigated for prior complaints or lost work time and claims due to body stressors.

Example: Your factory was designed and built in 1950. The workers were mainly younger males who had returned from the war, reasonably physically fit, and met the current "average" male profile of 5'10" tall, 180 pounds. Your workforce profile has changed over the years, and there are more older workers and more female workers in the factory. The paper punch machine is now operated by a 5'3" Hispanic man. The operation requires a twenty-five-pound bale of paper to be lifted off a pallet on the floor onto a roller tray 48" high and slid into place in the punch. The worker has lost work on several occasions because of back pain. The physical hazard here is identified as the lifting process of the heavy bales of paper and the height onto which the bales are lifted. In redesigning this work station to eliminate the hazard, the safety manger recommends placing the pallets of paper on a scissor jack that can be adjusted by the worker to bring the pallet up level with the roller trays. The bales can then be slid off the pallet without lifting.

NIOSH (2007) recommends that the

best way to reduce WMSDs is to use the principles of ergonomics to redesign tools, equipment, materials, or work processes. Simple changes can make a big difference. Using ergonomic ideas to improve tools, equipment, and jobs reduces workers' contact with those factors that can result in injury. When ergonomic changes are introduced into the workplace or job site, they should always be accompanied by worker training on how to use the new methods and equipment, and how to work safely.

NIOSH (2007) continues

Many ergonomics experts recommend that employers develop their own ergonomics programs to analyze risk factors at the worksite and find solutions. These programs may operate as part of the site's health and safety program or may be separate. An ergonomics program can be a valuable way to reduce injuries, improve worker morale, and lower workers' compensation costs. Often, these programs can also increase productivity.

There may be a particularly urgent need for an ergonomics program at your site if:

- Injury records or workers' compensation claims show excessive hand, arm, and shoulder problems; low back pain; or carpal tunnel syndrome.
- Workers often say that some tasks are causing aches, pains, or soreness, especially if these symptoms do not go away after a night's rest.
- There are jobs on the site that require forceful actions, movements that are repeated over and over, heavy lifting, overhead lifting, use of vibrating equipment, or awkward positions such as raising arms, bending over, or kneeling.
- Other businesses similar to yours have high rates of work-related musculoskeletal disorders.
- Trade magazines or insurance publications in your industry frequently cover these disorders.

Effective ergonomics programs have included the following elements:

- Employer commitment of time, personnel, and resources
- Someone in charge of the program who is authorized to make decisions and institute change
- Active employee involvement in identifying problems and finding solutions
- A clearly defined administrative structure (such as a committee)
- A system to identify and analyze risk factors
- A system to research, obtain, and implement solutions such as new equipment
- Worker and management training
- Medical care for injured workers
- Maintaining good injury records
- Regular evaluation of the program's effectiveness.

Training

Ergonomic issues in the workplace also require training of the workers to ensure they understand proper lifting techniques, use of back belts (or not!), proper posture when using computers, and reporting concerns. Let it be noted that NIOSH studies have found back belts to be useful in preventing back injuries but may give a false sense of security.

Ensure training is provided by individuals who have experience with the work processes and ergonomic issues. An OSHA 2008 publication states that

training prepares employees for active participation in the ergonomics process, including identifying potential problems, implementing solutions, and evaluating the process. Effective training includes:

- Proper use of equipment, tools, and machine controls;
- Good work practices, including proper lifting techniques;
- Awareness of work tasks that may lead to pain or injury;
- Recognition of MSDs [musculoskeletal disorders] and their early indications;
- Addressing early indications of MSDs before serious injury develops; and . . .
- Procedures for reporting work-related injuries and illnesses as required by OSHA's injury and illness recording and reporting regulation (29 CFR 1904).

Employees will benefit from orientation and hands-on training received prior to starting tasks with potential ergonomic risk factors. Employees should also be notified of workplace changes, instructed on using new equipment, and notified of new work procedures.

Controls

To eliminate and reduce the occurrence of WMSD injuries, there must be ergonomically conscious changes to equipment, work practices, and procedures. These ergonomic improvements will also help control costs and reduce employee turnover. Additionally, "These changes may also increase employee productivity and efficiency because they eliminate unnecessary movements and reduce heavy manual work. OSHA recommends that employers use engineering controls, where feasible, as the preferred method of dealing with ergonomic issues" (OSHA, 2008). Examples of ergonomic solutions to injuries caused by awkward positioning, repetitive motion, and vibrating equipment can be alleviated by:

- placing lights on an adjustable pole to avoid awkward work positioning
- placing devices in easy-to-each areas
- providing lightweight platforms for easier overhead access
- allowing the use of wheeled or motorized carts

- providing turntables and rotating tabletops that allow objects to be easily turned, rotated, and positioned
- making antivibration gloves, tools, and mats available to affected workers
- reducing the amount of time the worker has to perform the task.

Summary

It is the task of the safety and health professional to fit the job to the worker, not the worker to the job. Workspaces need to be designed ergonomically to prevent MSDs in employees. Watch the worker and the process, and identify hazards and workplace stressors. Where appropriate, redesign tasks, processes, tools, or materials.

Review Questions

1. What is ergonomics?
2. List some common MSDs.
3. What are some physical tasks that increase the chance of WMSDs?
4. What is the best way to reduce the occurrence of WMSDs?
5. Which elements may be included in an effective ergonomics program?
6. Effective training for employees may include which elements?

8

Electrical Safety

FAILURE TO ESTABLISH OR USE SAFETY PRACTICES for electrical equipment can result in property damage and serious injuries or fatalities. Nearly three hundred workers are killed each year from contact with electrical current or as the result of injuries caused by fires and explosions related to electrical accidents. A NIOSH study showed that during the period from 1992 through 2001, there were 44,363 electrical-related injuries involving days away from work. The number of nonfatal electrical shock injuries was 27,262, while 17,101 injuries were caused by electric arc flash burns.

Electrical energy consists of voltage (volts), current (amperes or amps), and resistance (ohms). The primary factor for injury from electrical sources is the amperage or current density. Severity of electrical shock is determined by

- the amount of current that flows through the victim (the body is a great conductor!)
- the length of time the body receives current (how long the victim is in contact)
- parts of the body involved (current flowing from source through body to ground).

As little as 100 milliamps of 60 Hz alternating current can be fatal (well below that in an average household). Above 12 milliamps, the average individual cannot release his or her grip on a handheld current source. High-voltage energy sources found in industrial settings are extremely hazardous.

- Most household electricity is 120 to 240 volts and 15–20 amps.
- Industrial settings have regular 120, 240, and high-voltage equipment, defined by the NEC as anything over 600 volts.
- Tasers and stun guns use 20,000 to 150,000 volts of electricity but only 2–3 milliamps of current—a nonfatal way to incapacitate.

The hazards of electricity at the power system supply level include electrical blast or flash, electrocution, short circuits, overloads, ground faults, and electrical fires.

A person's main resistance to current flow is skin surface. Dry skin has a fairly high resistance to the passage of electricity. Moisture dramatically lowers the skin's resistance. Once electricity breaches the skin barrier, the body offers little resistance to current flow. There are four main types of injuries: electrocution (fatal), electric shock, burns, and falls. These injuries can happen in various ways:

- direct contact with electrical energy
- when electricity arcs (jumps) through a gas (such as air) to a person who is grounded (that would provide an alternative route to the ground for electricity)
- thermal burns, including flash burns from heat generated by an electric arc, and flame burns from materials that catch on fire from heating or ignition by electrical currents. High-voltage contact burns can burn internal tissues while leaving only very small injuries on the outside of the skin.
- muscle contractions, or a startle reaction, can cause a person to fall from a ladder, scaffold, or aerial bucket. The fall can cause serious injuries.

Electrical hazards are best dealt with through *design*. This includes safe installations in accordance with *NFPA 70E: Standard for Electrical Safety in the Workplace* (2009a), the NEC, and local electrical codes.

- The NEC is the minimum standard but provides the single most comprehensive document dealing with installation and use of electrical equipment.
- In most states, it has been adopted as *law* for all new electrical construction and installations.
- Because it is a minimum standard, many states have developed more stringent codes.

Uniform enforcement for electrical safety–related work practices and policies are covered under OSHA regulations, 29 CFR 1910.331 through 29 CFR 1910.335.

Workers at risk of electrical injuries include utility, industrial, and commercial electrical personnel involved in hands-on roles or maintenance planning, such as engineers, technicians, electricians, linemen, supervisors, and other personnel who work on or near energized and de-energized electrical equipment and systems.

Management should determine the specific hazards in any particular equipment or its location before selecting and installing electrical equipment. The job analysis (JA) should identify electrical hazards and recommend controls. Electrical equipment and fixtures should be installed in controlled or isolated areas and have fail-safe devices, a means to completely de-energize the equipment, and guards to protect workers and others nearby. Electrical hazards are controlled to protect workers from shock and to protect equipment from damage. Listed are several safety and control devices and recommendations to take into consideration.

- Major equipment, such as generators and transformers are installed, if possible, in areas of limited access. Barriers must be installed if vehicular traffic passes nearby.
- All equipment must have a way of being positively de-energized. This can be a plug or a switch, fuse, or breaker. A secondary disconnect besides the regular switch on the machine ensures safety.
- All knobs, dials, handles, and buttons on controls, switches, drawers, and meters should be made of insulating materials.
- Fuses and circuit breakers are designed (hazard control through design!) to burn out or trip at a predetermined current.
- Control panels and operating boards are designed and laid out in a logical manner (emergency switches and cutoffs are painted *red* and installed in plain sight with unobstructed access).
- Ground fault interrupter circuits (GFIC) detect very small amounts of current leakage to ground potential. These devices can protect equipment and people by de-energizing the equipment before the hazard reaches a severe magnitude.
- Electrical equipment and motors must be protected from moisture, excessive dust, and flammable or combustible atmospheres.
- Grounding of equipment, including small cord-operated hand-tools, can prevent shocks and electrocutions and is the most important safety feature of any electrical equipment or system. All extension cords should be equipped with a three-prong plug. Bonding is used to ensure major parts of equipment are linked to provide a continuous path to the ground, such as bolting or welding equipment with good metal-to-metal contact.

- "High voltage danger" signs, CPR instruction signs, special precautions, and emergency procedures should be posted as required around electrical equipment.
- Fixed guards, security fences, or other barriers must be in place where workers may come into contact with wiring or electrical components. Warning devices should be used as necessary, such as lights, on/off signs, and audible signals.
- Electricians may need personal protective equipment (PPE), such as safety glasses or goggles, insulating gloves, insulating mats, arc suppression blankets, arc-flash clothing, arc-flash gloves, flame-resistant work clothes, voltage detectors, electrical safety shoes, and other safety tools and accessories, to safely work on circuits that may need to stay energized or may have capacitors.

Make sure that workers are thoroughly trained to check their electrical equipment for safe operation, to report any abnormal conditions, to have equipment tested regularly, and to observe all safety regulations and practices to operate specialized equipment. Be aware: the most frequently cited electrical safety hazards during OSHA inspections are:

- overloaded circuits
- frayed cords or disabled grounding plugs (cutoff ground prong or using "cheater" plug)
- missing covers on breaker panels, receptacles, connection boxes, and control panels
- jury-rigged electrical connections and bypasses.

Prior to inspections, electrical equipment should be de-energized whenever possible and the systems locked out or tagged to prevent accidental startup while the inspection or tests are being conducted. Management needs to know which equipment must be worked on while it is energized; administrative requirements may be applied to notify supervisors and get permission to work on this equipment. Only trained and specifically authorized employees may work on energized circuits.

All components of electrical equipment must be well maintained. Only trained and experienced electricians should make repairs to electrical circuits and apparatus. Electronic equipment may be sensitive to static electricity and surges, and may have capacitors that can be shock hazards. Only authorized, trained personnel should be working on electronic equipment.

A facility's safety program should include thorough training for all employees who work with electrical and electronic equipment or who operate

Figure 8.1. Electrical panel with a warning sign.

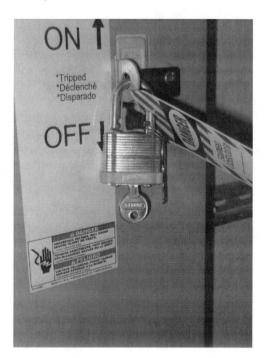

Figure 8.2. Locked and tagged electrical box.

electrical systems. Supervisors should know about existing and possible hazards, and all employees need to know CPR and rescue and other emergency procedures.

Non-equipment electrical hazards include exposure to lightning when working outdoors or near equipment susceptible to lightning strikes (towers). Personnel who routinely work outdoors in areas subject to thunderstorms need to be trained in precautions and CPR, and supervisors must be educated on how to avoid those hazards.

Summary

Because practically all members of the workforce are exposed to electrical energy during the performance of their daily duties and electrocutions occur to workers in various job categories, it is the safety professional's duty to control exposures. This can be done through training, implementation of a lockout/tagout (LOTO) program, signs and labeling, engineering out the hazards, and the proper use of PPE.

Review Questions

1. How is severity of electrical shock determined?
2. Discuss the physical hazards associated with electricity.
3. List several safety and control devices used to protect workers and equipment.
4. Which OSHA regulations cover electrical hazards?
5. Give an example of a non-equipment electrical hazard.

9

Fire Safety and Thermal Stressors

Fire Protection

FIRE PROTECTION INCLUDES PROCEDURES for preventing, detecting, and extinguishing fires to protect employees and property and to ensure continued operations. To accomplish these goals, companies must develop formal, written fire protection programs. The primary purpose of such programs is to prevent fires and train employees in proper response procedures should there be a fire. Employees should know their roles in detecting a fire and in transmitting an alarm, evacuating a building, or, if so assigned, confining and extinguishing the fire. Fire protection programs should enable companies to reduce hazards significantly through *prevention*.

Fire protection engineering is a highly developed engineering specialization. Achieving the most efficient fire protection system requires the involvement of architects, interior designers, urban planners, building contractors, electrical and structural engineers, fire detection system manufacturers, building safety engineers, and local fire departments. As a first step in fire prevention, every establishment should:

- set up a system of periodic fire inspections for every operation
- ensure that proper fire-extinguishing equipment is on hand and in good working order
- establish the inspection schedule, determine report routing, and maintain a complete list of inspected items
- set up regular fire drills for all personnel.

When a fire breaks out, good communication is vital to alert workers to the emergency and to mobilize fire protection forces. Companies should have coded fire alarms with complete backup systems. Cooling burning materials, removing oxygen from the fire, inhibiting the chemical chain reaction, and removing fuel can control fires. The objective of these methods is to interrupt the chain reaction in a fire. Fire-extinguishing agents help control fires in one or more of these ways.

The National Fire Protection Agency (NFPA, 2009b) has developed four classifications for fires:

1. Class A fires occur in ordinary paper, textiles, or wood materials and can be extinguished by water or dry chemical agents.
2. Class B fires occur in the vapor–air mixture over the surface of flammable or combustible liquids and are extinguished by foam or applying dry chemicals.
3. Class C fires occur in or near energized electrical equipment and can be put out by de-energizing and using dry chemical agents or carbon dioxide.
4. Class D fires occur in combustible metals and require special techniques and extinguishing agents and equipment to put out.

Fire protection measures are effective only if they are based on a proper analysis and evaluation of the risk of fire. Part of the job analysis (JA) should be to identify processes or areas that are potential sources of fires.

Example: During the JA, one of the daily requirements noted was to wipe down the rollers on the printing press at the end of each shift with a rag soaked in cleaning solvent. The cleaning solvent is classified as a flammable with a flash point around 50°F. The manufacturer requires that particular solvent to be used or it voids the machine warranty, so substituting a nonflammable liquid is not an option. Besides listing personal protective equipment (PPE) the worker will need for handling the solvent-soaked rags, you document several fire hazards—storage of the open solvent can under the operating machine, no extinguisher in the vicinity, no signs posted to warn about smoking or sparks, and a pile of solvent-soaked rags in the corner by the operator's mat.

Question: What is the most serious of these fire hazards?

Companies must organize a systematic fire hazard survey of all aspects of a facility and operations. Each operation, area, and the building itself must be evaluated for the fire hazard, preventive measures, response equipment on hand, and emergency exits.

In designing fire safety into a building, management must consider fire-fighting access to a building's interior, ventilation, connections for sprinklers and standpipes, traffic and transportation routes, and fire department and water access to the site. Design elements must be as fire resistant as possible and help to minimize fire hazards. Construction methods can help to confine fires and control smoke through proper design of stairways, firewalls, fire doors, separate units, ventilation ducts, physical barriers, and fire exits. Companies must act to eliminate the causes of industrial fires by using only approved equipment, establishing safe work practices, and enforcing good housekeeping procedures. Workers should be trained to spot unsafe or hazardous conditions and report them immediately. There are four construction methods designed to reduce the hazard by helping to confine fires:

1. stair enclosures
2. firewalls
3. fire doors
4. dividing a building into smaller units.

Fire detection must be part of every fire protection system. Its two main tasks involve (1) giving an early warning to enable building occupants to escape and (2) starting extinguishing procedures. Means of fire detection can be through a human observer; automatic sprinklers; smoke, flame, or heat detectors; or, best of all, a combination of these.

Plants or buildings should be equipped with fire alarm signal systems that clearly communicate to all personnel where the fire is located and that summon appropriate firefighting units. This includes accommodation for disabled workers, such as sight or hearing impaired, with audible and visual alarms. Employees must be trained to respond properly to alarm signals. Spacing, location, and maintenance of fire detectors depend on the type of building, its operations, and its materials.

Portable fire extinguishers are listed as class A, B, C, D, or a combination of A, B, and C, depending on the type of fire they are designed to extinguish. A recognized testing lab must approve this equipment, and it must be located in accessible area, clearly marked as to class and type of fire, and easily operated by workers.

Sprinklers and water spray systems come in many varieties, depending on the type of building, the operations performed in it, and the materials used.

- All systems require a reliable water supply of ample capacity and pressure.

- Automatic sprinklers are the most common and effective of all fixed fire-extinguishing systems and can serve as fire alarms as well as fire protection.
- Sprinkler systems and their water supplies must be inspected and tested regularly to ensure they function properly.

Fire hydrants, fire hoses, and hose nozzles should be available for immediate use but only by trained employees or responders. Hydrants are particularly effective in large plants where parts of the plant are far away from public fire hydrants or when no public hydrants are available. Fire hoses and nozzles should be inspected and maintained in good repair, and workers should be trained in their proper use during emergencies. Special fire hazards may require special fire-extinguishing or control agents other than water.

- These systems are usually installed to supplement, not replace, automatic sprinklers and other fixed or portable fire protection equipment.
- They must be engineered to fit the particular circumstances of a particular hazard.
- These special agents include foam systems, dry chemical piped systems, steam systems, and inert gas systems.

Figure 9.1. Closed flammable liquid storage with placarding.

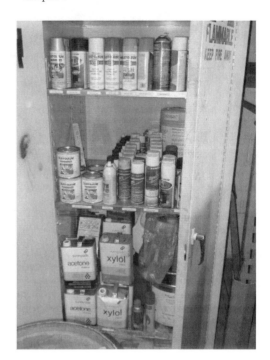

Figure 9.2. Open flammable liquid storage with placarding.

Foam systems are the most extensively used of these special agents. Chemical or mechanical means can generate foam. The principal types of foams available are low expansion for fighting class B fires and high expansion for fighting class A fires.

Buildings or shops where flammable, reactive, or combustible materials are used or stored should be posted with NFPA identification signs on the exterior of the building. This provides information to responders as to what materials they may encounter in responding to a fire or emergency. In the diamond-shaped symbol of the NFPA identification system, the three categories of hazards are identified for each material—health, flammability, and reactivity—and any special instructions. The numbers range from 0 to 4, with 4 being the most severe hazard and 0 indicating relatively no hazard. Special instructions, such as the example W, means "use no water," provide other details.

Federal law (Department of Transportation) also requires identification of hazardous materials using a similar diamond-shaped label as a placard whenever rail, air, water, and public highways ship the materials. The numbers on the placard correspond to the material, the hazard, and the response.

Flammable and Combustible Liquids

Hazardous materials can present two types of hazards:

1. chemical hazard—acute or chronic effects due to toxicity, corrosive properties, carcinogen, and so on.
2. physical hazard—fire hazard, sudden release of pressure, or reactivity.

A material can fall into more than one category, of course—a flammable liquid may also pose a skin irritation hazard or be carcinogenic, like benzene. The most common physical hazard of hazardous materials is flammability.

- A flammable liquid is any liquid having a flash point below 100° F (37.8° C).
- A combustible liquid is any liquid having a flash point between 100° F (37.8° C) and 200° F (93.3° C).
- Flammable gases form a flammable mixture with air at a concentration of 13 percent or less, such as acetylene and propane.
- Flammable aerosols are aerosols that will project a flame back to the valve or more than eighteen inches out from the valve (usually from the propellant).
- Flammable solids are likely to cause a fire, can be ignited readily, and will burn vigorously and persistently, such as magnesium metal.
- Oxidizers can initiate or promote combustion and release oxygen when in contact with organic materials. The gas production can present an explosion hazard (chlorine, fluorine, hydrogen peroxide).
- Pyrophoric materials can ignite spontaneously in air and at temperatures below 130° F (white phosphorous, linseed oil–soaked rags, etc.).

Major fires with significant property damage and loss of life have resulted from improper or unsafe use of flammable and combustible liquids. These fires can result in:

- loss of production and, potentially, customers
- destruction of business records and costs for reconstruction
- direct costs for losses not fully insured or depreciated
- increased insurance premium
- water and smoke damage to equipment and property.

Flammable and combustible liquids can present a physical hazard during transport, storage, handling, and disposal of the liquids. A fire can occur at any point but most frequently occurs with storage of the liquid.

The inherent hazard of these liquids is determined largely by (1) flash point, (2) concentration of vapors in the air, (3) risk of ignition at or above the flash point, and (4) amount of vapor present. Examples of commonly used industrial flammables and combustibles are:

Fuels: gasoline, diesel, and bunker oil
Solvents: cleaning solvents and paint thinners
Paints: oil-based paints, enamels, and varnishes
Oils: lubricating oil, greases, and cutting fluids.

Workers should always avoid exposing large surface areas of flammable or combustible liquids to the air to avoid creating a serious fire or explosion risk. This includes leaving tanks uncovered or accidental spills of materials. Other general safety measures include:

- prohibiting smoking around or near these liquids
- shielding the liquids from static electricity by bonding and grounding tanks and storage cans
- using only spark-resistant tools when working with these liquids.

Flammable and combustible materials should remain in their original containers, if possible, or dispensed into safety cans for use. Once used, contaminated debris or rags must be stored in safety cans or containers until final disposal.

Only trained employees should load or unload trucks or vessels used to transport flammable or combustible liquids. Some materials are dispensed by tank car or truck into large storage tanks or bulk containers.

- Underground storage tanks should be protected against overhead traffic, built on a firm foundation, and shielded against corrosion and leakage.
- Above-ground tanks must be constructed an approved distance away from property lines, public ways, and nearby buildings and be surrounded by containment dikes.
- All tanks should be equipped with proper fire-extinguishing systems.

Storage and mixing rooms inside a facility should be isolated and protected as much as possible to guard against external fire hazards. Outside storage lockers should be located or built at a distance from the main plant whenever possible and must conform to NFPA regulations.

Cleaning tanks that contained flammable and combustible liquids is extremely hazardous work and requires highly skilled and trained employees.

- Workers must wear protective equipment and understand the proper work and medical procedures to follow (*no smoking*, no keys or exposed metal on the body, etc.).
- Small tanks and containers can be cleaned by steam and made temporarily safe by means of an inert gas.
- Tanks to be abandoned must be thoroughly cleaned, then dismantled and removed from the premises.

Proper disposal of flammable and combustible liquids is an important part of handling these materials safely.

- If recycling or recovery of these liquids is impossible, they should be burned in an Environmental Protection Agency (EPA)–approved incinerator or given to a disposal contractor.
- Items, such as rags soaked with a flammable or combustible liquid, must be stored in NFPA-approved metal containers, which are emptied daily as directed by local EPA standards.

Because flammable and combustible liquids have many uses in industry, workers must know how to guard against fire hazards, prevent unintentional incidents, and protect their health when using flammable and combustible liquids.

Thermal Hazards

Thermal stressors or hazards can ignite fires or explosions, damage equipment, or injure workers. Examples of hot surfaces or thermal hazards include:

- electric heaters or hot plates
- operating engines or compressors, especially the exhaust manifolds
- boiler and furnace surfaces, stacks, and chimneys
- steam radiators, lines, and equipment
- burning cigarettes
- metals heated by friction, such as brake drums and bearings
- surfaces heated by radiation from the sun or from fires
- metals being welded
- radiation [infrared (IR), electromagnetic, solar, microwave (MW), radiant, or ultraviolet (UV)]
- laser beams
- metallic sparks.

Workers can also be injured by excessively *cold* surfaces or liquids, such as cryogenic operations, refrigeration piping, and extremely cold metal surfaces. Physical damage to equipment that can result from thermal hazards include warping of metal parts out of tolerance, burning of lubricating fluids causing metal-to-metal parts to seize, melting of hoses and belts, and fire damage.

The physical injury to workers that can result from thermal hazards includes:

- first-, second-, or third-degree burns to the skin from contact with hot metals, hot surfaces, slag, hot water or steam, or sparks
- first-, second-, or third-degree burns to the skin from exposure to UV or other radiation
- destruction of the eye tissue or retina from lasers or UV radiation
- heat stroke, heat stress, or heat disorders from exposure to hot and humid environments
- inhalation of byproducts of heated oils and lubricants
- internal damage from exposure to MW
- frostbite from exposure to extremely cold surfaces or cryogenic liquids.

Thermal hazards should be identified during the JA. Hot and cold surfaces that present a burn hazard should be guarded from any possible skin contact. Radiation sources must be shielded. Where the hazard cannot be eliminated, protective equipment, such as welder's helmets for UV and spark protection, must be specified. Workers in areas or conditions where there are extremes of temperature must be properly outfitted with PPE. The workers must be provided training on the hazards and symptoms of exposure, and their supervisors trained in management of personnel working in temperature extremes.

Boilers and Unfired Pressure Vessels

Boilers and unfired pressure vessels have many potential hazards in common that must be controlled by safety devices and safe work practices. Boilers are heated, pressurized vessels used to produce steam from water. They are most commonly found as propulsion units in vessels, steam-producing units for manufacturing processes, and home-heating furnaces. They can become a hazard if they are over-pressurized and explode. Unfired pressure vessels are normally unheated ("unfired") containers that contain gasses or liquids under pressure.

- If exposed to external heat sources, the gases or liquids in these vessels can expand and create an explosion hazard. For example, an aerosol can thrown into a fire will explode from the pressure built up when the contents are heated.

- If over-pressurized, an unfired pressure vessel can explode or components on the system, at the weakest point, can fail. For example, the pressure gauge installed on the top of a propane tank can be blown off the tank if over-pressurized.

Management must see to it that the design, fabrication, testing, and installation of boilers and unfired pressure vessels comply strictly with all federal, state, and local codes.

- The American Society of Mechanical Engineers (ASME) and the National Board of Boiler and Pressure Vessel (NB) inspectors have established guidelines for inspecting this equipment.
- Workers should receive training to operate equipment according to safety standards and safe work practices, especially maintenance personnel.
- Operators must be carefully selected and thoroughly trained to operate pressure vessels safely, to understand emergency procedures, and to use a checklist to inspect and maintain equipment. Supervisors must closely monitor all operations.

All operations involving boilers and unfired pressure vessels must be identified during the JA. All components associated with these vessels must be included in the JA, such as maintaining valves, gauges, and safety devices. Lockout and tagout (LOTO) are critical elements to safe maintenance and repair of pressurized units. Safety devices on all pressure vessels will help prevent accidents and injuries. Workers must be sure that these devices are properly installed, tested, and maintained. Safety devices include:

safety valves: ASME/NB-rated and -stamped safety valves. Spring-loaded valves are commonly used safety devices for pressure vessels. These are also called "pop" valves because they pop fully open when a preset pressure is exceeded. They are normally used for gases, steam, and vapors.

relief valves: These valves open when the upstream pressure exceeds some predetermined level but only opens in proportion to the amount of overpressure. These are used on home hot water heaters.

rupture disks: This is a relatively flat metal piece that is designed to burst at a particular pressure. A frangible disk may not clog as easily as a spring-loaded safety valve and is easily and inexpensively replaced. A rupture disk must function within ±5 percent of its specified bursting pressure at a specific temperature.

vacuum breakers: These work similarly to a spring-loaded or relief valve, except that the atmospheric pressure outside the vessel, being higher than the inside of the vessel, forces it open. These are commonly used on water systems; when the pressure drops in the water piping, the vacuum breaker opens to equalize the pressure.

Physical hazards from failures of pressurized vessels may include the following:

- Discharges from safety valves must be conducted in locations where the operation of a safety or relief valve does not cause an even greater hazard (where is the chemical liquid or gas going to go?). This is especially true if the discharge is a flammable, toxic, or corrosive liquid or vapor.
- Compressed air and gas cylinders must be handled such that the valves on the ends of the tanks are protected. Protective caps must be replaced after use. A dropped cylinder where the valve breaks off can become a torpedo traveling at fifty feet per second.
- Whipping of unsecured flexible hoses can also cause severe injury and damage. This applies to water hoses, air or compressed gas hoses, or steam hoses. Hoses should be replaced with rigid lines wherever possible.
- Over-pressurization until failure can occur when relief devices are not provided. For example, an over-pressurized truck tire will fail catastrophically because it lacks any relief device.

Other injuries can occur when personnel begin work on equipment but fail to relieve the pressure in the system before opening the system. Workers have been injured by searching for system leaks using bare hands—high-pressure leaks can easily amputate a finger (the "broom" test!).

Physical hazards also exist for negative pressure systems (vacuums). Excessive negative pressure or condensation of vapors in a vessel can cause it to collapse, which may release material or pop off metal components, such as gauges.

Summary

It is the employer's responsibility to train workers about fire hazards in the workplace and the procedures they should follow in a fire-emergency situation. Workers need to know how to prevent fire emergencies from occurring, what to do in the event of a fire emergency, how to use firefighting equipment and whom to contact. The key to fire safety is training.

Review Questions

1. List the steps to setting up a fire-prevention program.
2. Name and define the four classifications of fires.
3. What are the four construction methods designed to reduce the hazard of fires?
4. When can combustible and flammable liquids present a physical hazard?
5. Give three examples of thermal hazards.
6. How can thermal stressors be controlled?

10

Hand and Portable Power Tools

HAND AND PORTABLE POWER TOOLS are used extensively in every type of industrial operation, both in production and maintenance. They are also found extensively in home workshops. Although manufacturers have tried to make hand and power tools as safe as possible (such as using rechargeable batteries instead of power cords), frequent injuries occur from use, or misuse, of these tools. In the workplace, many disabling injuries such as burns, cuts, eye trauma, amputations, and cumulative trauma disorders (CTDs) can be attributed to hand-tool usage.

Manual tools include various hammers, drills, wrenches, handsaws, chisels, gardening tools, screwdrivers, and so on. The majority of injuries result from misuse of the tool—wrong tool for the job (chiseling with a screwdriver), using a tool incorrectly (hammering with a wrench), or forcing a tool that is over- or undersized for the job (a wrench slipping off a nut when hit with a hammer).

Portable power tools have a power source, such as:

• electric (cord or battery)
• pneumatic (compressed air)
• gasoline (internal combustion engines)
• hydraulic (compressed hydraulic fluid)
• powder-actuated (gun powder actuates and drives nails).

Injuries resulting from misuse of power tools are usually more serious and can be fatal—for example, electrocutions, driving nails through body parts, flying debris that damages eyes, and amputations from power saws. Home use of tools is the responsibility of the user. Manufacturers must list all safety hazards and precautions on packaging. Safety in the occupational use of hand and powered tools is the responsibility of the employer.

Proper selection, use, care, and supervision of hand and power tools can prevent abuse of these tools and eliminate or reduce employee injuries. Management must select proper hand and power tools and change tools only after careful consideration. Control used to prevent injury during power-tool usage can include adding personal protective equipment (PPE), job rotation, or other adjustments.

Six safety practices workplaces should enforce and their employees should follow include:

1. usage of proper protective equipment
2. selection of the right tool for the job
3. upkeep and maintenance; keep tools in good condition—inspect before use
4. making sure power tools are properly grounded
5. training workers; tools must be used correctly
6. storage of tools in a safe place.

Due to the pilferage of and maintenance requirements for some tools, many companies set up tool-issue areas, or centralized tool control. Using this method, tools can be issued and signed for by individuals who are responsible for keeping them in a locked toolbox or locker when not in use. Some companies may prefer a centralized tool issue, where tools are checked out daily and returned at the end of the shift or job. The advantage of having centralized tool control in an industrial setting is that it ensures uniform inspection and maintenance by a trained employee. Tools can be issued with the proper PPE by a trained employee. Safety precautions can be reviewed with the worker checking out the tool before each use.

OSHA regulations cover hand and portable power tools. Regulation 29 CFR 1910.242(a) provides the general requirements. Each employer shall be responsible for the safe condition of tools and equipment used by employees, including tools and equipment that may be furnished by employees. In other words, if an employee decides to bring in his own hammer and wrenches to use on the job, then it is the employer's responsibility for the safety of the tools and worker. Hand and powered portable tools used in shipyards and marine terminals are covered under 29 CFR 1915 and 1917. Construction standards, under 29 CFR 1926, provide safety regulations for a wide variety of tools.

Summary

The key to safe hand and portable power-tool use is ensuring that workers use the right tool for the job, and they use it correctly. It is the employer's responsibility to provide the proper tools and training to employees. If selection criteria or worker training are overlooked, injuries are bound to occur.

Review Questions

1. List common injuries associated with hand and power-tool use.
2. What is the main cause of injury associated with hand and power-tool use?
3. How can hand and power-tool injuries be prevented?
4. Discuss the safety practices and controls that can aid in the prevention of hand and power-tool–related injuries.
5. Explain the benefit of a central tool-issue room.

11

Woodworking

CARPENTRY SHOPS AND THE WOODWORKING trades offer their own set of unique hazards. OSHA regulations in 29 CFR 1910.213 cover the design, use, personal protective equipment, and precautions for woodworking machinery. Machines used in woodworking are dangerous, particularly when used improperly or without proper safeguards. Workers operating woodworking equipment suffer the following common injuries: laceration, amputation, and blindness. Wood dust and the chemicals used in finishing are health hazards, and workers in this industry can suffer from skin and respiratory diseases. Common woodworking shops contain nonportable, installed, shop equipment. Following is a list of the most-used machinery:

- circular, crosscut, and ripsaws
- overhead swing and straight-line pull cutoff saws
- radial saws
- band saws
- jigsaws
- jointers
- shapers
- power-feed planers/moulders
- lathes
- sanders
- routers
- tenoning machines
- boring/drilling/mortising machines.

OSHA's *Guide for Protecting Workers from Woodworking Hazards* (1999) states that the principal hazards of woodworking can be classified as either safety or health hazards. Safety hazards can cause immediate injury to a worker. For example, if not properly grounded, the metal framework of a circular saw could become energized and possibly electrocute an employee. Or, if a worker's hands were to contact a saw blade, he or she could have one or more fingers cut off. The physical hazards of woodworking include:

- machine hazards
  - point of operation
  - rotary and reciprocating movements
  - in-running nip points (pinch points)
- kickbacks
- flying chips, material
- tool projection
- fire and explosion hazards
- electrical hazards
- noise
- vibration
- wood dust—explosive hazard and biological hazard (carcinogens).

There are also chemical hazards from exposure to coatings, finishings, adhesives, and solvent vapors in woodworking. Most injuries from woodworking machines occur at the point of operation. The point of operation is the place where work is performed on the material. This is where the stock is cut, shaped, bored, or formed. Most woodworking machines use a cutting and/ or shearing action.

Employees can be injured if their hands get too close to a blade, particularly when working on small pieces of stock. The size of the piece dictates that the operator's hand be close to the blade. Accidents can occur when stock unexpectedly moves or when a worker's hand slips. Injuries can occur if:

- stock gets stuck in a blade and actually pulls the operator's hands into the machine
- the machine or its guard is not properly adjusted or maintained (An improperly adjusted radial saw, for example, might not return to its starting position after making a cut.)
- the machine has controls that are not recessed or remote, and the equipment is accidentally started, then a worker's hands may be caught at the point of operation

- contact occurs during machine repair or cleaning (Lockout/tagout procedures must be followed.)
- an employee reaches in to clean a saw or remove a piece of wood after the saw has been turned off but is still coasting or idling. Also, saw blades often move so fast that it can be difficult to determine whether they are moving. This is especially a problem under fluorescent lighting (strobe effect).

Rotating or reciprocating machinery also presents serious physical hazards—all machines operate by rotating or reciprocating motion or by a combination of these motions. Rotary cutting and shearing mechanisms, rotating wood stock, flywheels, shaft ends, and spindles all rotate.

- Rotating action is hazardous regardless of the speed, size, or surface finish of the moving part.
- Rotating parts and shafts, such as stock projecting from the chuck of a lathe, can catch hair or clothing and draw the operator in. This can seriously mangle or crush the operator.
- Rotating parts and stock can also force an arm or hand into a dangerous position, breaking bones and lacerating or severing a hand or other parts of a limb.
- Bolts, projecting keys, or screws on rotating parts increase the danger of getting caught by the rotary part. Operators can also be struck by a projecting bolt or key.
- In-running nip points (or pinch points) are a special danger arising from rotating or reciprocating parts. They occur whenever machine parts move toward each other or when one part moves past a stationary object. Parts of the body may be caught between or drawn into the nip point and be crushed, mangled, or severed.

Example: "Router Operator Killed by Flying Tool"

A 32-year-old experienced woodworker was fatally injured at work while operating an over-arm router. The worker was making custom rosettes when a steel-tool knife was propelled from the rosette cutter. The knife penetrated a Plexiglas shield and then penetrated and exited his chest. The knife ricocheted off a wall before landing. The knife, measuring approximately 1 5/8 inches square, was part of a cutter-head assembly that had been previously used on a drill press at much lower cutting speeds. It was custom designed for the drill press, not for the router, which is run at much higher speeds. The knife was held in the cutter head by flat shims and set screws; the screws could not counteract the centrifugal forces generated by the high-speed rotation (Massachusetts Department of Public Health, 1997).

Another hazard is kickbacks. These occur when a saw seizes the stock and hurls it back at the operator. This can happen when the stock twists and binds against the side of the blades or is caught in the teeth. A blade that is not sharpened or is set at an incorrect height can cause kickbacks. Also poor-quality lumber (in other words, frozen lumber or lumber with many knots or foreign objects, such as nails) can also result in kickbacks. Hazards due to kickbacks can occur when there is a lack of safeguards, such as spreaders, antikickback fingers, and gauge or rip fences. It is best to use a crosscutting method against the grain of the wood to prevent kickbacks. Other than kickbacks from wood, employees may be exposed to flying splinters and chips that are flung by the cutting action of woodworking equipment.

Another common hazard is from tool projection (unbalanced cutter heads). Many pieces of woodworking equipment—such as routers, shapers, and molders—employ rotating cutter heads with multiple knives.

- Cutter heads that are not properly adjusted, are poorly mounted, or have broken knives can become unbalanced. Balance is critical for keeping knives secured to a rapidly moving cutter head.
- The centrifugal forces on an unbalanced cutter head can fling the knives from the tool and severely or fatally injure the operator or other nearby personnel.
- Using the wrong tool on a cutter head or using a tool at a higher speed than it was designed to operate can cause tool breakage and projection.

Hazard Controls

It cannot be stated enough times: The preferred way to control hazards is through engineering or work practice controls. When these controls are not possible or do not provide adequate protection, PPE must be provided as a supplement. Employers must institute all feasible engineering and work practice controls to eliminate or reduce hazards before using PPE to protect employees.

Engineering controls involve physically changing the machine or work environment to prevent employee exposure to the potential hazard. Examples are using a guard on a machine, or using local exhaust ventilation to remove dust and other contaminants at the source. Engineering controls used on this type of machinery include:

- machine guards
- location and distance

- automatic feed and ejection systems
- interlocks and controls to prevent accidental startup
- counterweights to pull a hazard away from the user (such as on saws)
- placement of the controls/two-handed controls.

Work practice controls involve removing employees from exposure to the potential hazard by changing the way they do their jobs. For example, workers should always use push sticks to guide short or narrow pieces of stock through saws. Using a push stick allows saw operators to keep their hands at a safe distance from the saw blades. Common administrative and work practice controls for machinery include the following:

- Use appropriate equipment for the job. Workers can be seriously injured if they do not use the correct equipment for a job. Use machines only for work within the rated capacity specified by the machine manufacturer. Use the correct tools on a given machine.
- Train workers on machine use and allow only trained and authorized workers to operate and maintain the equipment. Employees should be able to demonstrate their ability to run the machine with all safety precautions and mechanisms in place.
- Frequently inspect equipment and guards. Ensure that: (1) the operator and machine are equipped with the safety accessories suitable for the hazards of the job, (2) the machine and safety equipment are in proper working condition, and (3) the machine operator is properly trained. Document the inspections and keep the records.
- Use equipment only when guards are in place and in working order. A worker should not be allowed to operate a piece of woodworking equipment if the guard or any other safety device, return device, spreader, antikickback fingers apparatus, guard on in-running rolls, or gauge or rip fence is not functioning properly.
- Provide employees with push sticks or other hand-tools so that their hands are away from the point of operation when they work on small pieces of stock. A push stick is a strip of wood or block with a notch cut into one end that is used to push short or narrow lengths of material through saws.
- Use a brush or stick to clean sawdust and scrap from a machine. Never allow employees to clean a saw with their hands or while the machine is running.
- Provide regular preventive maintenance. Regularly clean and maintain woodworking equipment and guards. Ensure that blades are in good condition. Knives and cutting heads must be kept sharp, properly

adjusted, and secured. Sharpening blades prevents kickback. You must also remove any cracked or damaged blades from service. Keep circular saw blades round and balanced.
- Never leave a machine unattended in the "on" position. Make sure that workers know never to leave a machine that has been turned off but is still coasting.
- Maintain proper housekeeping. Workers have been injured by tripping and then falling onto the blades of saws. You must keep floors and aisles in good repair and free from debris, dust, protruding nails, unevenness, or other tripping hazards. Do not use compressed air to blow away chips and debris. Make sure you have a nonslip floor.
- Do not allow workers to wear loose clothing or long hair. Loose clothing or long hair can be easily caught up in rotating parts.
- Never saw freehand. Always hold the stock against a gauge or fence. Freehand sawing increases the likelihood of an operator's hands coming in contact with the blade.

PPE encompasses a wide variety of devices and garments designed to protect workers from injuries. Examples include respirators, goggles, safety shields, hard hats, gloves, earmuffs, and earplugs. Common woodworking PPE includes:

- hard hats
- safety glasses, goggles, and face shields
- gloves (including chemically protective gloves)
- padded kickback aprons; vests; and arm, groin, and leg guards
- lower-back supports
- steel-shank, steel-toed safety shoes with slip-resistant soles
- earplugs and earmuffs
- particulate-resistant and/or chemically resistant overalls
- respirators.

Fire Prevention in the Woodworking Shop

Woodworking facilities are inherently prone to fires and explosions. Woodworking shops are filled with a large amount of wood, wood products, sawdust, and flammable materials like paints, oil finishes, adhesives, solvents, and liquid propane. These materials are literally fuel for the fire. Of particular concern is the abundant production of sawdust, which can ignite and burn far more easily than whole pieces of lumber. Sanders, routers, and shapers are known for producing large amounts of fine dust. Very fine wood dust is especially hazardous. It can accumulate on rafters and other structural components and in unexpected spots all around the facility, far from the point of generation. Other ignition sources can arise from faulty electrical wiring, sparking tools, propellant-actuated tools, and employee smoking.

Preventing the buildup of dust is one of the key means for controlling fire and explosion hazards. The principal engineering control technology for dust is exhaust ventilation. The primary work practice control is good housekeeping. Dust collection is best achieved if it is done at the source—at the point of operation of the equipment. For many pieces of equipment, local ventilation systems consisting of well-designed ducts and vacuum hoods can collect most of the dust generated before it even reaches the operator. The local exhaust systems must have a suitable collector. Dust collection systems must be located outside the building, unless the exceptions described in NFPA standards are met. Very fine dust that manages to escape point-of-source collection can be captured from above by general exhaust points located along the ceiling. These control technologies are effective for most equipment.

Good housekeeping extends to periodic hand cleaning of your entire facility, as some dust will escape from even the best exhaust systems and will eventually accumulate on rafters and other out-of-the-way spots. Also, it is extremely important to inspect and clean your exhaust ventilation system on a regular basis to maintain maximum efficiency. Never permit blow-down of accumulated dust with compressed air. Blowing dust with compressed air will create the very type of dust cloud that presents the greatest explosion hazard.

It is also imperative to ensure the proper use and storage of flammable materials, such as paints, finishes, adhesives, and solvents. Make sure to segregate tasks particularly prone to fire and explosion hazards, such as spray painting, welding, and use of powder-actuated nail guns. As previously discussed, train employees to recognize, avoid, and correct potentially hazardous conditions and behaviors. Train employees so that they are acquainted with the special equipment and aspects of building design related to dealing with fires and explosions. Control all ignition sources by using electrical systems rated for the projected use and protected by appropriate circuit breakers, grounding all equipment prone to accumulating static electrical charges, grounding entire buildings against the possibility of lightning strikes, and controlling and banning smoking in and around the workplace.

Summary

Woodworking and carpentry shops come with their own unique set of hazards, ranging from dangerous machinery to fire hazards. The good news is many of the occupational hazards present in woodworking shops can be controlled, thus lessening or eliminating the hazard. Workplaces and their employees need to remain cognizant and vigilant to prevent serious injuries and accidents from occurring.

Review Questions

1. List several hazards associated with woodworking operations.
2. During woodworking operations, where do the majority of injuries occur? Why?
3. What type of controls can be put in place to prevent injury in a woodworking shop?
4. Why must the buildup of dust be kept to a minimum? And how is dust controlled?
5. Give examples of fire prevention practices that can be helpful in a woodworking shop.

12

Metalworking Machinery

METALWORKING MACHINERY INVOLVES LARGE, stationary machines installed in workshops, with point-of-operation hazards similar to woodworking shops. Metalworking machinery is used for cutting, shaping, boring, bending, or forming various bar, plate, or sheet metals. These machines are commonly used on any type of metal or alloys. The main difference between woodworking and metalworking machinery is the inherent hazards of the stock that is being worked. For example, unlike wood chips, metal chips or shavings can be very sharp and cause serious lacerations and eye damage.

Metalworking machinery includes:

- metal cutoff saws
- drill presses
- sheet metal brakes (benders)
- sheet metal shears
- lathes
- rolls (rollers to flatten or curve sheet metal)
- mills and borers
- sanders/deburring machines
- crimpers/benders.

Metalworking machinery can process a range of metal parts from a micro level, such as surgical tools or jewelry, up to huge, multiton gears and parts.

Physical Hazards

As with any other industrial machinery, the physical hazards of metalworking include:

- machine hazards
  - point of operation
  - rotary and reciprocating movements
  - in-running nip points (pinch points)
- flying metal chips, material
- tool projection
- cutting fluid toxicity
- electrical hazards
- noise
- vibration.

Unsafe work practices or incorrect procedures most often cause injuries on machine tools. Proper safeguarding on machines, good housekeeping in the work area, and good work habits help to reduce injuries and accidents. Common metalworking injuries include:

- amputations and crushing injuries
- lacerations and bruises
- loss of sight
- cumulative trauma disorders (CTDs)
- dermatitis (from various cutting fluids and oils).

Common injuries that can occur during metalworking operations include:

- being struck by insecurely clamped work or by tools left on or near a revolving table
- catching clothing, hair, jewelry, or rags for wiping in revolving parts
- falling against revolving work
- calipering or checking work while the machine is in motion
- allowing sharp metal turnings to build up on the table
- removing turnings by hand
- not washing after contacting cutting fluids or getting the fluid in the eyes.

Controls

Engineering controls involve physically changing the machine or work environment to prevent employee exposure to the potential hazard. Engineering controls used on this type of machinery include:

- machine guards
- location and distance
- automatic feed and ejection systems
- interlocks and controls to prevent accidental startup
- counterweights to pull a hazard away from the user (such as on sheet metal brakes)
- placement of the controls/two-handed controls.

Common work practice controls for machinery include the following:

- Use appropriate equipment for the job. For example, a worker must use the correct bit in a drill press for the hardness of the metal to be drilled.
- Train workers on machine use, and allow only trained and authorized workers to operate and maintain the equipment.
- Frequently inspect equipment and guards.
- Use equipment only when guards are in place and in working order.
- Use a brush or stick to clean metal shavings and scrap from a machine.
- Provide regular preventive maintenance.
- Never leave a machine unattended in the "on" position.
- Maintain proper housekeeping.
- Do not allow workers to wear loose clothing or long hair.

Personal protective equipment may be necessary where the operator is in direct vicinity or contact with the machine or stock. Common metalworking PPE includes:

- safety glasses, goggles, and face shields
- gloves (including chemically protective gloves when handling cutting fluids)
- steel-shank, steel-toed safety shoes with slip-resistant soles
- earplugs and earmuffs.

Metalworking Fluids

One physical and chemical hazard of metalworking besides the machinery itself is from metalworking fluids (MWFs). MWFs are a range of oils and other liquids used to cool and lubricate the piece of metal being worked, especially during cutting and milling. MWFs are classified as either:

- "straight oils" or "neat oils" (not meant to be diluted with water and may contain highly refined petroleum, animal, marine, vegetable, or synthetic oils)

- soluble oil (highly refined petroleum oils and emulsifiers)
- semisynthetic fluids
- synthetic fluids (which may include detergent-like components).

All MWF classes may contain additives, such as stabilizers, biocides, dispersants, dyes, and odorants. When MWFs are used, a primary concern is the presence of contaminants that encourage the growth of bacteria and fungi. Also, there is a potential for oils to be heated high enough that the cutting tool works on a metal workpiece to form polynuclear aromatic hydrocarbons (PAHs). While MWFs are used by hundreds of thousands of workers safely, problems can develop when good hygiene practices are not followed or when fluids are not properly managed or maintained. Major health concerns of improperly managed fluids or when good hygiene practices are not followed include:

- skin irritation
- allergic contact dermatitis
- irritation of the eyes, nose, and throat
- breathing difficulties, such as bronchitis and asthma.

Although rare, some workers have contracted hypersensitivity pneumonitis (HP) from improperly managed fluids. HP is an allergic-type reaction in the lungs that may be caused by exposure to certain microbial products. HP is marked by chills, fever, shortness of breath, and a deep cough—similar to a cold that will not go away (OSHA, 2007c).

Cold Forming of Metals

According to OSHA, cold forming is the process of using press brakes, rollers, or other methods to shape steel into desired cross-sections at room temperature. The safety of power presses, brakes, and rollers depends on:

- adequate safeguarding of the point of operation
- proper training of operators
- enforcing safe working practices.

When determining point-of-operation safeguards, consider all hazards that may crush, cut, punch, sever, or otherwise injure workers. Safeguarding devices control access to the point of operation and can be either press controlling, operator controlling, or a combination of the two. One newer type of control used as a safeguard device is a presence-sensing device. This device is designed, constructed, and arranged to create a sensing field or area and to deactivate the clutch control when an operator's hand or any other body part is detected in the area.

Hot Working of Metals

According to OSHA, hot work refers to any activity involving riveting, welding, burning, powder-actuated tools, or similar fire-producing operations. Grinding, drilling, abrasive blasting, or similar spark-producing operations are also considered hot work except when such operations are isolated physically from any atmosphere containing more than 10 percent of the lower explosive limit of a flammable or combustible substance. Hot working of metals involves melting or increasing the temperature of a metal in order to forge or reshape it.

Hazards of Hot Work

The inherent physical hazards of hot working metals is from the extremely high temperatures to which the metal is heated or melted, ranging from 327° C (620° F) for low-temperature melting-point alloys, such as zinc or tin, to over 3,600° C (6,500° F) for high-temperature melting-point alloys, such as steel- or nickel-based alloys. Common injuries and illnesses associated with hot working metals include:

- burns from hot surfaces, slag, sparks, and ultraviolet and radiant heat
- noise
- heat stress
- flash burns to the eyes and increased incidence of cataracts
- musculoskeletal disorders and cumulative trauma disorders.

Melting scrap metals can create fumes from old paints, lubricants, and coatings and lead, nickel, or chromium additives. Sparks or molten metal from hot work, when combined with oxygen and a fuel source, can start a fire. Hot work is only allowed in areas that are free of fire hazards or where fire hazards have been controlled by physical isolation, fire watches, or other positive means. Employers must maintain hazard-free conditions in the space while hot work is being performed. Other hazardous conditions can also result in fires. Potential ignition sources include combustible materials that may cause or contribute to the spread of fire, such as cigarettes, matches, heat guns, heat-producing chemical reactions, electric shocks, and improper use of heating devices.

Controls

As it is reiterated in every section of the book, worker protection from hazards begins with engineering controls and administrative controls. These controls are used to decrease hazards to workers. In the instance of hot metalworking, PPE is critical to operators who are working in and around hot metals. Engineering controls for hot work may include:

- local and general ventilation
- machine guarding
- shields and barriers
- use of less-toxic substances.

Administrative controls for hot work may include:

- hot work–permitting system
- safety training (Do not perform hot work where flammable vapors or combustible materials exist. Work and equipment should be relocated outside of the hazardous areas, when possible.)
- signs and labels ("no smoking," "hot work area")
- lockout/tagout (LOTO) program and training
- confined space entry permit and training
- the availability of adequate fire-watch/fire-protection equipment.

Recommended PPE for hot metalworking includes:

- leather boots
- leather gloves or gauntlets
- safety glasses and/or welder's helmet
- hearing protection
- respirators
- apron, jacket or cape, leggings, and spats made of leather, aluminized glass fabrics, synthetic fabrics, or treated wool
- a full-coverage hot-work coverall with hood of aluminized glass fabric.

Performing Hot Work in a Confined Space

Take notice if performing hot work in a confined space for accumulation of toxic gases. In confined spaces, a hazardous atmosphere may exist. The circumstances could create an oxygen-deficient (atmospheric concentration of less than 19.5 percent) or oxygen-enriched (atmospheric concentration of more than 23.5 percent) environment.

Make sure to ventilate toxic metal fumes mechanically if entering a confined space, such as inside a mud tank, water tank, oil tank, hopper, sump, pit, or cellar. OSHA requires the use of a written permit system to document authorization to enter, the work to be performed, and the results of the gas monitoring where there is a potential for toxic, flammable, or oxygen-deficient atmosphere. Both a hot-work and confined-entry permit may be required for welding, cutting, or brazing within a confined space.

Summary

Most injuries incurred during metalworking involve unsafe work practices. It is the duty of the employer to provide proper job hazard analysis (JHA), employee training, and equipment that is in safe working order. Common workplace injuries involving metalworking machinery can be prevented with good housekeeping and maintenance, employee training (safe work practices), and the use of machine guards or other engineering controls that eliminate or reduce the hazard.

Review Questions

1. List the physical hazards associated with metalworking machinery.
2. Discuss the controls associated with metalworking machinery.
3. What are some common injuries incurred during metalworking operations?
4. Give examples of physical hazards associated with hot working of metals.
5. What needs to be noted about performing hot work in a confined space?

13

Welding Operations

WELDING, CUTTING, AND BRAZING ARE HAZARDOUS activities that pose a unique combination of both safety and health risks to more than 500,000 workers in a wide variety of industries. The risk from fatal injuries alone is more than four deaths per one thousand workers over a working lifetime (OSHA, 2007d). The basic safety standards for welding and cutting are found in OSHA 29 CFR 1910.251 through 255 and provide general precautions and personal protective equipment requirements. Additional OSHA regulations for welding and cutting operation safety in shipyards and marine terminals are covered under 29 CFR 1915 and 1917, and construction standards, under 29 CFR 1926.

The Bureau of Labor Statistics (BLS, 2009) states that

welding is the most common way of permanently joining metal parts. In this process, heat is applied to metal pieces, melting and fusing them to form a permanent bond. Because of its strength, welding is used in shipbuilding, automobile manufacturing and repair, aerospace applications, and thousands of other manufacturing activities. Welding also is used to join beams in the construction of buildings, bridges, and other structures and to join pipes in pipelines, power plants, and refineries.

There are several variations of arc welding, including shielded metal arc welding (SMAW) or stick welding, metal inert gas, tungsten inert gas, and plasma arc welding. Arc welding uses electrical currents to create heat and bond metals together, but there are over one hundred different processes that a welder can employ. The type of weld used is normally determined by the types of metals being joined and the conditions under which the welding is to take place. Steel, for instance, can be welded more easily than titanium. Some of these processes involve manually using a rod and heat to join metals.

Like welders, brazing workers use molten metal to join two pieces of metal. The metal added during the soldering and brazing process has a melting point lower than that of the piece, so only the added metal is melted, not the piece. Brazing uses metals with a higher melting point. Brazing often is used to connect copper plumbing pipes and thinner metals that the higher temperatures of welding would warp. Brazing also can be used to apply coatings to parts to reduce wear and protect against corrosion (BLS, 2009).

Thermal cutting involves two processes, oxygen and arc cutting, that cut the metal by melting. Oxygen cutting is usually performed on carbon, manganese, and low-chromium-content steels. The metal is exposed to heat, and oxygen causes it to melt and oxidize. Another type of thermal cutting, called flame cutting, uses fuel gas or a combination of gases, such as acetylene, hydrogen, natural gas, or propane. These gases when heated cause the metal to vaporize and separate. Arc cutting encompasses several types of processes, including plasma arc cutting, carbon arc cutting, and oxygen arc cutting, to name a few. Arc cutting is accomplished by using the heat of the arc to melt the metal, causing a cut or separation. It is usually used with nonferrous metal, stainless steels, or chromium- or tungsten-containing steels (DHHS, 1988).

Hazards and Controls

Welding without the proper precautions can be a dangerous and unhealthy practice. However, with the use of new technology and proper protection, risks of injury and death associated with welding can be greatly reduced. Physical injury can result from burns because many common welding procedures involve an open electric arc or flame. To prevent them, welders wear PPE in the form of heavy leather gloves and protective long-sleeve jackets to avoid exposure to extreme heat and flames. Ultraviolet radiation is generated by the electric arc in the welding process. Skin exposure to UV radiation can result in severe burns, in many cases without prior warning. UV radiation can also damage the lens of the eye. Many arc welders are aware of the condition known as "arc-eye," a sensation of sand in the eyes. This condition is caused by excessive eye exposure to UV radiation. Exposure to UV rays may also increase the skin effects of some industrial chemicals (coal, tar, and cresol compounds, for example). Exposure to infrared (IR) radiation produced by the electric arc and other flame-cutting equipment may heat the skin surface and the tissues immediately below the surface. Except for this effect, which can progress to thermal burns in some situations, IR radiation is not dangerous to welders. Most welders protect themselves from IR (and UV) radiation with a welder's helmet (or glasses) and protective clothing.

Exposure of the human eye to intense visible light can produce adaptation, pupillary reflex, and shading of the eyes. Such actions are protective mechanisms to prevent excessive light from being focused on the retina. In the arc-welding process, eye exposure to intense visible light is prevented for the most part by the welder's helmet. However, some individuals have sustained retinal damage due to careless viewing of the arc. At no time should the arc be observed without eye protection (OSHA, 1996c).

Goggles and welding helmets with dark faceplates are worn to prevent this exposure, and in recent years, new helmet models have been produced that feature a faceplate that self-darkens upon exposure to high amounts of UV light. To protect bystanders, translucent welding curtains often surround the welding area. These curtains, made of a polyvinyl chloride plastic film, shield nearby workers from exposure to the UV light from the electric arc but should not be used to replace the filter glass used in helmets.

Welders are also often exposed to dangerous gases and particulate matter. Processes like flux-cored arc welding and shielded-metal arc welding produce smoke containing particles of various types of oxides, which in some cases can lead to medical conditions like metal fume fever. The size of the particles in question tends to influence the toxicity of the fumes, with smaller particles presenting a greater danger. Many processes produce fumes and various gases, most commonly carbon dioxide, ozone, and heavy metals, which can prove dangerous without proper ventilation and training.

Mechanical ventilation consists of either general mechanical ventilation systems or local exhaust systems. Ventilation is deemed adequate if it is of sufficient capacity and so arranged as to remove fumes and smoke at the source and keep their concentration in the breathing zone within safe limits as defined in subpart D of part 1926 of Occupational Health and Environmental Controls. Contaminated air exhausted from a working space shall be discharged clear of the source of intake air, and all air replacing that withdrawn shall be clean and respirable.

Special considerations and regulations must be followed when welding, cutting, and heating in confined spaces. Except where air-line respirators are required or allowed as described later, adequate mechanical ventilation meeting the requirements described earlier shall be provided whenever welding, cutting, or heating is performed in a confined space. When sufficient ventilation cannot be obtained without blocking the means of access, employees in the confined space shall be protected by air-line respirators in accordance with the requirements of subpart E of part 1926 of Personal Protective and Life Saving Equipment. An employee on the outside of the confined space shall be assigned to maintain communication with those working within it and to aid them in an emergency. Where a welder must enter a confined space through a small opening, means shall be provided for quickly removing him in case of emergency. When safety belts and lifelines are used for this purpose, they shall be so attached to the welder's body that his body cannot be jammed in a small-exit opening. An attendant with a preplanned rescue procedure shall be stationed outside to observe the welder at all times and be capable of putting rescue operations into effect (OSHA, 1996c).

Fire Prevention and Compressed Gas Cylinders

The use of compressed gases and flames in many welding processes poses an explosion and fire risk, so some common precautions include limiting the amount of oxygen in the air and keeping combustible materials away from the workplace. Compressed gas cylinders containing oxygen, acetylene, aron, helium, nitrogen, and other compressed gases may be moved, stored, or used

Figure 13.1.   Compressed gas cylinders.

in welding areas. Specific care must be taken with compressed gas cylinders, which includes the following:

- Many welding processes take place off-site from a protected welding shop. Cylinders must be safely transported to those sites using carts.
- Cylinders not in use should have their valves turned off, hoses removed, and safety caps replaced. Empty cylinders should be marked as MT.
- Compressed gas cylinders are color coded (oxygen cylinders are always painted green, or have a green label, for example).
- Hoses used for compressed gases are often also color coded.
- Spare oxygen (accelerator) and acetylene (flammable gas) should not be stored together in bulk. Flammable gases are to be stored in fire-protected areas (alarms, sprinklers, etc.).

All welders and welder's helpers must be trained on use of the equipment, safety precautions, PPE, and emergency procedures. No welding, cutting or heating shall be done where the application of flammable paints, the presence of other flammable compounds, or heavy dust concentrations create a hazard. Suitable fire-extinguishing equipment shall be immediately available in the work area and shall be maintained in a state or readiness for instant use.

Welder's helpers are equipped with PPE and fire extinguishers and stand by where the welder is working to stop the welder if a fire occurs in adjacent areas. Special precautions and requirements pertain to welding operations underwater, in confined spaces, in tanks, and other hazardous areas (OSHA, 1996c).

Summary

Physical hazards associated with welding operations can be reduced and eliminated through the use of engineering and administrative controls and the use of appropriate PPE. Safety training is a must for welders and those working around welding hazards. Without the proper training, employees are very likely to become exposed to welding arcs, thus increasing the chance for injury.

Review Questions

1. Define welding. List several types of welding.
2. What are the differences among welding, cutting, and brazing?
3. Identify the physical hazards associated with welding operations.
4. How can these hazards be controlled?
5. What special fire-prevention precautions need to be taken?

14

Vehicular and Mobile Equipment Safety

Powered Industrial Trucks

According to OSHA, each year tens of thousands of injuries related to powered industrial trucks (PITs), or forklifts, occur in U.S. workplaces. Many employees are injured when lift trucks are inadvertently driven off loading docks, fall between docks and an unsecured trailer, are struck by a lift truck, or fall while on elevated pallets and tines. Most incidents also involve property damage, including damage to overhead sprinklers, racking, pipes, walls, and machinery. Unfortunately, most employee injuries and property damage can be attributed to lack of safe operating procedures, lack of safety rule enforcement, and insufficient or inadequate training.

PITs are used for material handling when the loads are too large, heavy, or bulky to be handled manually. For example, moving a pallet load of thirty cartons by forklift is more efficient and involves less manual labor than hand-carrying each box from location to location. Six types of PITs include:

1. lift trucks, such as forklifts
2. straddle lifts
3. crane trucks
4. tractors and trailers
5. motorized hand trucks
6. automated guided vehicles (AGVs).

Figure 14.1.   Two thousand–pound electric PIT.

While more efficient than manually handling materials, industrial trucks pose a physical hazard like any other vehicle. To use this equipment safely, the company should:

- evaluate and establish a safe work environment where the truck will be used—design the plant to accommodate moving vehicles
- select the proper vehicle for the job and ensure that it meets or exceeds safety requirements
- screen and train the drivers
- set up an inspection and maintenance program for the trucks
- review the safety programs, accidents, and trends.

Safe Work Environment

When deciding to use powered trucks in a plant, warehouse, or store, ask and answer the following questions:

- Will it be used outside in inclement weather, ice, snow, rain, and so on?
- Are there flammable liquids, vapors, or explosive dusts to consider?
- Are the areas where it is driven equipped with guardrails, ramps, adequate clearances, smooth pavement, or crushed stone? Are there mud, railroad tracks, unguarded loading docks, or other hazards present?

- Is there adequate ventilation for gasoline, diesel, liquefied petroleum gas (LPG), or propane-powered trucks if used inside a building?
- Is there a safe area for recharging electric battery–powered trucks?
- Are driving areas equipped with warning signs, alarms, corner mirrors, traffic signals, or other control devices?
- Is there adequate lighting around the truck operation area for the driver and nearby workers to see the truck's movements?

Selecting a Safe Powered Truck

The most commonly used powered trucks are forklifts or platform styles of lift trucks. The operator stands on the vehicle, sits on the vehicle, or walks alongside the vehicle. The most important consideration in selecting a safe powered truck is to match the vehicle with the intended use.

Example: A walking lift truck would not be selected to move loads long distances over rough walking surfaces. This would tempt the operator to unsafely ride the truck, increasing the risk for an accident. It also decreases efficiency.

Power source, operator position, or means of engaging the load will determine the type of powered industrial trucks purchased. Besides matching the truck to the job, other factors considered in purchasing trucks may include:

- worksite constraints (such as narrow aisles)
- operator comfort
- safety features, such as backup alarms, safety belts, belly switches, overhead protection cages, and wraparound seats.

The safety features and intended use must always be considered.

Example A: AGVs must have some means of stopping should someone step in front of them. Such trucks should be equipped with flexible bumpers that shut off power on contact.

Industrial trucks should not be used for any purpose other than the one for which they were designed. Horseplay is often encountered and has resulted in serious and fatal accidents.

Example: Two fairly young and inexperienced forklift drivers and their buddy were on lunch break at the warehouse where they had a summer job. At the end of their break, they decided to race back to their work areas, the loser to buy at happy hour that evening. The buddy said he would ride on the forks of the one truck so he did not have to walk back to the job site. The forklifts do not move very fast but fast enough that when the one with the rider spun around a corner, the rider was thrown from his footing and hit by the second forklift, killing him.

Lift trucks not specifically equipped with personnel platforms are not authorized to lift people, and workers are not authorized to ride on any part of the truck where they are not the operator.

Powered trucks must be selected to handle the greatest capacity load anticipated. If the load does not fit on the forks in a balanced position, then the load has to be placed on a pallet, or a platform truck should be used to lift the load. Loads cannot be lifted using only one fork, and two forklifts cannot share one load. Lift trucks cannot be modified subsequent to purchase. (It's never a good idea to weld a rack on the back of one forklift to accommodate an ice chest!)

OSHA's Powered Industrial Truck Standard 29 CFR 1910.178 requires that all PITs purchased and used after February 15, 1972, must meet the design and construction specifications of ANSI B56.1-1969. Any truck used to perform tasks beyond its design and rated capacity is not only unsafe—it violates OSHA regulations.

If the truck is to be used in a hazardous location, the truck must be selected in accordance with OSHA-approved designations, such as diesel powered, completely enclosed motors, gasoline powered, nonsparking electrical powered motors, and so on.

Selecting and Training Drivers

Drivers should be medically prescreened to verify that they are fit to safely handle a PIT. Screening should include evaluation of their vision, depth perception, and hearing. Training programs should center on company policies, operating conditions, and types of trucks used. Both classroom and driving instruction are used, as well as final certification or licensing exams. Some companies make their own stricter policies for certifications and refresher training. Management should maintain records of each employee's driving performance and must maintain records of the driver's training.

Operators of industrial trucks can prevent accidents by using the same safe-driving techniques they employ on the highways. Operators are responsible for the care of trucks and should never leave a truck unattended, park in an aisle or doorway, idle engines for too long, or ignore mechanical problems. Operators should be aware of the basic differences between lift trucks and automobiles or highway trucks, which include the following:

- Lift trucks are generally steered by the rear wheels.
- Lift trucks steer more easily when loaded rather than empty.
- Lift trucks are driven in the reverse direction as often as in the forward.
- Lift trucks are often steered with one hand, the other hand being used to operate the controls.

Inspection and Maintenance

Preoperational checklists should be used prior to the operation of a PIT by the operator. Damaged vehicles should be tagged out of service until repaired and inspected for safe operation. The manufacturer will provide a maintenance schedule and list of preventive, periodic checks for each type of truck. Service life maintenance and repair records must be maintained for every vehicle. These records may be critical when reviewing accidents or when looking at safety trends. Make sure that operators who are responsible for changing propane tanks, refueling, or changing or charging the batteries are trained in the precautions and personal protective equipment required for these operations.

Industrial Truck Safety Program

Design for hazard control applies to selecting the safest truck for the job, designating the areas where that truck can be used, and having safety involved in any new purchases and changes to the procedures or location of processes. Accident and near-accident records should be reviewed periodically for trends. Unsafe drivers may need retraining or be removed from the job. Lack of maintenance and timely repairs may contribute to accidents, such as brake failures.

Accidents involving PITs are frequently serious and may result in a fatality or material damage to the company's equipment or goods. The most frequent accidents involving industrial trucks include:

- workers being pinned between the truck and a stationary object (wall)
- material falling off the fork, damaging material or injuring the operator

- shocks and acid burns from handling large forklift batteries
- overturned trucks due to overloading or uneven load
- trucks backing off loading docks
- workers nearby being run over or struck on the legs by lowered forks.

Haulage and Off-Road Equipment

Earth movers, dump trucks, road-surfacing equipment, backhoes, bulldozers, graders, and other construction vehicles require specialized training and pose physical hazards due to their size, terrain on which they are used, and environmental conditions. OSHA regulations that cover construction sites, including use of heavy equipment, are found in 29 CFR 1926—Construction Industry.

It is imperative that companies maintain safety features and equipment, train operators, and train repair and maintenance personnel to prevent accidental damage stemming from heavy equipment operation. Only qualified drivers should operate heavy-duty equipment. All heavy-duty vehicles must be equipped with substantial guards, shields, canopies, and grilles to protect the operator.

When workers are transported, supervisors must make sure that employees know safety procedures for riding and getting on and off the vehicles. Likewise, towing requires safe practices to prevent accidents in the coupling and uncoupling of motorized equipment. Operators are responsible for the safe working of their equipment, inspecting all parts and reporting any problems, and the safety of those working in areas around the machines. Companies must carefully train employees in safe practices for maintenance and repair of powered equipment and in safe operating procedures. All employees should observe good housekeeping practices. Operators and other workers on a construction or worksite must be aware of the special hazards associated with heavy-duty equipment.

Supervisors must ensure that employees observe safety practices at all times.

- Safety briefings should be provided for *all* workers at the site.
- Special attention must be paid to environmental conditions—rain, mud, flood areas, thunderstorms, and underground or overhead utilities.
- Work near active roadways requires special attention and caution.

Besides safety devices on the equipment, operators, signal personnel, and coworkers may be required to wear PPE:

- hard hats
- safety shoes
- safety glasses/safety sunglasses
- reflective or brightly colored safety vests
- earplugs
- drinking water
- communication devices.

Orange traffic cones, flags, and signs may also be required at the job site.

Hoisting and Conveying Equipment

Hoisting equipment is used to move loads over variable paths within a restricted area.

- It is generally used when there is insufficient or intermittent flow volume, such that a conveyor cannot be justified.
- Hoisting equipment provides more flexibility in movement than a conveyor but less flexibility than a forklift.
- Usually, loads handled using hoists are more varied with respect to their shape and weight than those handled by conveyor.
- Most cranes use hoists for the vertical movement.

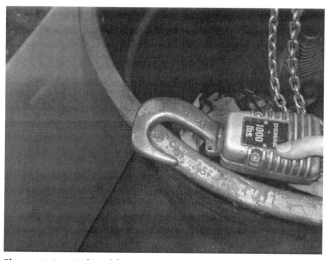

Figure 14.2. Hoist with a missing safety latch.

Types of cranes include monorails, jib cranes, derricks, tower and mobile cranes, and portable floor cranes. They all have specific uses according to their design, weight capacities, and movements. All have specific guidelines for safe operation and transport. Hoisting equipment must never be overloaded or used to transport people unless specifically designed for personnel.

Crane design and safe use is covered by OSHA under 29 CFR 1910. 179, Overhead and Gantry Cranes; 29 CFR 1910.180, Crawler Locomotive and Truck Cranes; 29 CFR 1917, Marine Terminals; and 29 CFR 1918 Long Shoring. These regulations cover design and construction of the equipment, weight testing and certifications, use limitations, operator training, maintenance, and inspections of the equipment. Operators should examine hoists regularly and repair or replace any damaged or malfunctioning part. The safety of personnel and equipment requires that:

- cranes have adequate safeguards to provide safe footing and accesses for the operator, to prevent injuries, and to limit the action of the crane arm and hoisting devices
- all hoisting ropes, slings, sheaves, drums, and other equipment is appropriate for the crane being used
- the operator's cab protects the operator against fire and weather, is well ventilated, contains ample control equipment, and allows for a clear view of signals
- aerial basket lifts are commonly used for working above ground
- manufacturer's operating and maintenance instructions are followed for safe handling of machine parts
- loads are lifted when directly under the hoist because angled loads can impose dangerous stresses on the equipment. If the load is not properly centered and tended, it can swing and cause injury or damage.

Conveyors use rollers, belts, chains, buckets, hoppers, and screws to move materials consistently from point to point. Unlike hoisting equipment, most material moved by conveyors is of a standard size and weight. Conveyors are used:

- when material is to be moved frequently between specific points
- to move materials over a fixed path
- when there is a sufficient flow volume to justify the fixed conveyor investment.

Conveyors

For safe operation of conveyors the loading points must be clearly marked, showing the safe load limit that can be moved, with safeguards used along the entire length. Operators must stand clear of moving conveyors, avoid pinch points and other areas where hands or fingers can get caught, and never attempt to repair a moving belt. During repair work, ensure that maintenance personnel lock out all power before working on a conveyor.

Elevators are conveyors used to move people and materials in a vertical manner. The code that governs the use and design of elevators is known as the Elevator Code, which is provided by ANSI and the American Society of Mechanical Engineers (ASME) A17.2, *Elevators, Escalators, and Moving Walk* (2007). The Elevator Code requires that:

- power elevators conform to the Elevator Code for electric-drive and hydraulic-drive elevators
- belt-driven and chain-driven machines never be installed
- interlocks and electric contacts for both passenger and freight elevators be direct-acting-mechanical and activated devices that cannot be inactivated
- companies establish a regular program of inspection, testing, and maintenance of all elevator parts according to city, state, and federal regulations
- companies select elevator operators carefully and establish proper working procedures, safety rules, and emergency procedures for operators to follow.

Escalators are conveyors for moving personnel. Requirements for escalators include easily accessible emergency stop buttons and sensing devices that will interrupt the power if the preset speed is exceeded or a tread chain breaks. The principal hazards on escalators and moving walks arise from their misuse by the public. The most often associated physical hazards pertaining to the use of hoists and conveyors typically arise from:

- being caught in moving equipment
- falling materials
- loads striking workers or other material causing injury or damage
- failures of equipment due to overloading or incorrect use.

Summary

The NIOSH (2001) investigations of forklift-related deaths indicate that many workers and employers may not be aware of the risks of operating or working near forklifts and are not following the procedures set forth in OSHA standards, consensus standards, or equipment manufacturer's guidelines.

Generally, reducing the risk of forklift incidents requires comprehensive worker training, systematic traffic management, a safe work environment, a safe forklift, and safe work practices.

Review Questions

1. List six types of PITs and the hazards associated with them.
2. What are some safety mistakes employees make during the operation of PITs?
3. Give examples of controls used to prevent workplace injury during the operation of PITs.
4. What are some other types of motorized equipment used to move loads? What safety hazards might they pose?
5. Which OSHA standards cover PITs and other haulage and off-road equipment?

15

Retail, Service, and Warehouse Facilities

PHYSICAL HAZARDS ARE NORMALLY associated with heavy industry, machines, tools, and industrial processes. Retail and service industries may have fewer severe physical hazards but have a higher percentage of workers. These industries include:

- office workers (white collar workers)
- hospital and health service industries
- home repair (electricians, plumbers, roofers, etc.)
- cashiers, store clerks, supervisors
- stockers in retail stores
- teachers and educators
- managers and supervisors
- bus drivers, taxi drivers, and so on
- city, state, and municipal workers
- warehouse workers (material handling)
- food service workers and wait staff.

There are over 500,000 lost–work-time cases per year costing $13 billion annually attributed to retail, service, and industries—mainly from musculo-skeletal disorders and cumulative trauma disorders. OSHA regulations that particularly apply to the service industry include the general duty clause, posting requirements, reporting and recordkeeping, means of egress, hazard communication, medical services and first aid, walking and working surfaces, lockout/tagout (LOTO), electrical installations and equipment, machine guarding, and powered equipment.

Hazards

Physical hazards in service, retail, and warehouse industries include:

- CTDs
- musculoskeletal stresses and strains
- exposures to dust, molds, chemicals, and extremes of temperature
- stress and violence
- road or traffic hazards
- slips, trips, and falls
- electrical shock (office equipment).

Controls

Good safety management in the service and retail industry is based on a strong corporate commitment to safety and its investment in time to develop specific guidelines and policies for training employees. This investment pays off through reduced injury rates, reduced workers' compensation problems, reduced regulatory fines, better-quality service, and better profit margins. Because the service industry is physically demanding, companies must make sure that employees are physically fit to do their work. They should also provide employees with proper training in safe work practices and procedures.

Companies in retail and distribution can manage their risks and losses by:

- training employees in proper incident reporting and analysis procedures
- introducing motor fleet safety programs
- conducting safety audits
- training workers in material handling
- providing all required personal protective equipment (PPE)
- introducing ergonomics to the workplace
- planning for emergencies and disasters
- ensuring greater security against crime and violence on the job.

Larger department or "big-box" stores have corporate-run programs for employee orientation, including safety. Enforcement is up to the local managers. Bulletin boards in common rooms, training rooms, or break rooms should be posted with various OSHA regulations and worker rights, how to report accidents, and who the company safety officials are and their contact information.

An employee manual should not only define the required level of safety but also give specific instructions. The following topics should be included in an orientation program:

- employee responsibilities and discipline
- safety rules
- accident and incident reporting
- general fire protection
- emergency procedures
- security alarms and inventory controls.

The high turnover in retail jobs complicates safety training and general orientation. The use of a standard program that can be read or the use of a computer-assisted training course provides a cost-effective method of training. Too often, safety training consists of one employee of limited experience telling the new guy what needs to be done and how to do it, and he or she may not mention safety or may even tell the new employee to do something that is unsafe.

Engineering controls for hazards in the retail and service industries include:

- ergonomically designed work stations or work areas
- material-handling equipment and aids
- antifatigue mats at standing work stations
- good general ventilation and temperature controls
- good general and task lighting.

PPE may be required, such as gloves and safety shoes, for material handlers.

Job analysis is important in these types of jobs to evaluate a work environment. These are usually considered low-priority, low-risk work environments but can result in more compensation claims than in an industrial plant.

The age of workers varies widely, from summer teens to postretirement part-timers. The JA must take into account the age and physical condition of the workers when evaluating the process or procedure. People who can no longer lift boxes in the warehouse may be reassigned to a computer terminal doing inventories but then suffer further back pain from a poorly designed work station or nonadjustable chair.

The JA even in office spaces must fully evaluate the work environment. While a clean, air-conditioned office space may at first glace seem like a nice, safe work environment, worker's compensation records and health complaints may reveal safety issues. Whether a worker is away from an office job or warehouse job, it still impacts company costs and production.

Transportation Safety Programs

Transportation safety includes on-road vehicles, railways, aircraft, and ships. The transportation applies to cargo, raw or finished materials, people, animals, or anything else that may be commercially transported.

On-Road Vehicle Safety

On-road vehicles (cars, trucks, tractor-trailers, delivery vans, etc.) owned and operated by a company, often referred to as a "fleet," have unique hazards. Besides routine on-road driving hazards, maneuvering trucks and trailers around loading docks, warehouses, ramps, and container cranes requires additional training and experience. Fleet, commercial, and tractor-trailer drivers are required to hold a commercial driver's license (CDL). Classroom and skills training are required for a CDL, as well as a physical examination and medical clearance.

Companies that run fleets, delivery trucks, hauling trucks, or tractor-trailers know that there are safety hazards that not only can cause death or injury to their drivers and damage to or loss of their vehicles but also affect the other vehicles and drivers their personnel may encounter on the road. The Department of Transportation (DOT), Federal Motor Carrier Safety Regulations (FMCSR) establish regulations for everything from the size and placement of handholds to climb up into trucks to the physical requirements to drive. For example, a driver cannot qualify for a CDL if he or she has diabetes, heart problems, poor eyesight, missing limbs, or epilepsy. The FMCSR also provides the basic requirements for a vehicle safety program. A vehicle safety program should include:

- a written safety policy
- a designated safety program manager
- efficient accident investigation and reporting systems
- a driver selection and training program
- a preventive vehicle maintenance program.

The safety and health professional or fleet manager is responsible for supervising the program and reporting on safety issues to top management.

A driver safety program should include a training program, collision prevention measures, reporting procedures, driver performance goals and incident reports, and a method for establishing competency and skill levels and collision/safety records for each driver. If the driver is also the delivery person, lifting and carrying safety training should be included.

The total cost of vehicle incidents almost always exceeds the amount recovered from the insurance company and includes direct and indirect expenses of collisions. The costs of vehicle collision prevention programs are more than justified when compared with potential losses related to collisions. A motor vehicle collision can be defined as any incident in which the vehicle comes in contact with another vehicle, person, object, or animal with resulting injury or property damage.

Hazards

The hazards of on-road vehicles include:

- road and highway accidents
- hazardous materials and cargos
- CTDs and MSDs
- drug and alcohol use.

Controls

Many on-road accidents can be prevented through driver training, enforcement of driving hours, and good maintenance of vehicles. Training must include proper lifting and manual material handling, hazardous material emergency procedures, and use of safety equipment while driving (safety belts, etc.).

The DOT mandates drug and alcohol testing of all employees who work in safety-sensitive jobs, such as operating motor vehicles. Employers must establish testing programs in compliance with DOT rules and regulations.

Railway Safety

The DOT has a Federal Railroad Administration that defines rail safety regulations and requirements. There are about five thousand train accidents per year, with an average of twenty railroad employee deaths per year. The number of railroad-related deaths by trespassers on tracks and vehicles crossing tracks averages five hundred per year. The Federal Railroad Administration employs more than 415 federal safety inspectors nationwide, operating out of eight regional offices. They specialize in five safety disciplines:

1. hazardous materials
2. motive power and equipment
3. operating practices (including drug and alcohol)
4. signal and train control
5. track and structures.

The inspectors conduct site-specific safety inspections of railroads and monitor their compliance with federally mandated safety standards. They also collect and analyze rail accident and incident data from the railroads and convert this information into meaningful statistical tables, charts, and reports.

Hazards

Railroad hazards exist for employees, such as:

• noise
• hazardous material exposure from tank cars and cargo
• CTDs and musculoskeletal issues
• environmental exposures (heat, cold, sun)
• crushing, amputation, and severe injuries from large moving railcars.

Controls

Railroad employee safety programs, as well as the involvement of established unions, have significantly reduced the number of fatalities and injuries among workers over the past fifty to one hundred years. Training, physical qualifications, and safety devices are key to a successful program. There are elaborate qualification programs for locomotion engineers, for example.

Flight Safety

The primary goal of a flight safety department is to enhance aviation safety by preventing incidents through comprehensive safety programs and investigation. Safety education can be used to inform workers and others of an organization's operations and provide other airlines with information to develop their own plans. To reduce aviation incidents, an organization must have a well-developed and practiced response plan that can be implemented immediately. The primary purpose of incident investigation is to provide the data necessary to prevent similar occurrences.

The Federal Aviation Administration (FAA) sets maintenance requirements, inspections, and pilot qualification criteria; conducts investigations; and establishes safety regulations for all private and commercial aircraft. While flying is one of the safest modes of transportation, in-flight accidents are usually catastrophic when they occur. Most of the aviation employee–related accidents occur with ground crews, baggage handlers, maintenance personnel, and refueling crews.

Hazards

Flight hazards may involve:

- noise
- hazardous material exposure from fuels
- CTDs and musculoskeletal issues
- environmental exposures (heat, cold, sun)
- crushing, amputation, and severe injuries from moving vehicles
- violence and terrorist activities.

Controls

Airline employee safety programs enforce FAA safety regulations. Engineering controls are becoming more common, such as modern innovations in baggage handling. Training, safety supervision, physical qualifications, and more automation has been used to reduce accidents. PPE is required for ground crews for noise, visibility, and head hazards.

Ships and Marine Terminals

Maritime industry hazards include marine terminals, at-sea hazards, and hazards within the ships themselves. The American Bureau of Shipping (ABS) and the U.S. Coast Guard (USCG) establish maritime safety programs, construction criteria, and emergency response programs. As with any transportation industry, training of employees, established safety programs, and safety inspection programs help prevent accidents and control hazards. OSHA's 29 CFR 1917 establishes extensive safety regulations for workers at marine terminals. The 29 CFR 1918 pertains to long-shoring, which includes working on tugboats and loading or off-loading of cargo ships.

Hazards

Hazards at marine terminals include:

- movement of large cargo containers
- powered industrial trucks
- noise
- extremes of temperature
- hazardous material handling
- entry into confined spaces and cargo holds
- biological hazards from cargo (food goods)
- cumulative trauma and musculoskeletal disorders
- cranes, winches, and conveyor movement
- slippery conditions
- pirates.

Merchant marine crew qualification, certifications, and regulations are enforced by the USCG. Safety of crew members and safety of life at sea (SOLAS) are also regulated by the USCG. This includes lifesaving, firefighting at sea, such safety equipment as life boats, and the design of the ships for safety, such as emergency escapes.

Controls

Control of safety hazards for the marine terminals, long-shoring, and on cargo ships is through inspections, enforcement of regulations, and engineering out physical hazards. For example, most engineering or propulsion plants on cargo ships are now fully automated. This is a long way from the hot and dangerous conditions when boilers were fed by stokers shoveling coal.

Automated Lines, Systems, or Processes

Automation is an engineering control for physical, chemical, and biological hazards, depending on the process. Automation and robotics are very effective controls for physical hazards.

The best-known form of the assembly line, the moving assembly line, was realized into practice by Ford Motor Company between 1908 and 1913. Their initial idea came from engineers observing "disassembly" plants—slaughterhouses where carcasses were butchered while they moved along a conveyor. Ford was the first company to build large factories around the assembly line concept.

Automation introduces physical hazards unique to workplace environments. In an assembly line, *everything is in motion!* Automated production requires special hazard precautions because of the nature of the machinery, the variety of human–machine interactions on the job, and the rapid development of automated technology.

Safety in automated manufacturing processes can be enhanced by

careful identification of hazards
development of appropriate strategies for hazard controls
worker training to interact with the workplace environment.

Hazards

Components of many automated manufacturing systems or cells include computerized numerically controlled machines, materials-handling and transport systems automated guided vehicles, and robots.

As with any man–machine interface, physical hazards include:

- noise
- machine motions (nip points, rotating shafts, moving conveyors, robots)
- ergonomic hazards
- eye hazards
- electrical shock.

Companies should provide visual and mechanical warning signs, awareness barriers, interlocks, and controls to protect workers from hazards of automated equipment or systems.

Workers most at risk in assembly line environments are maintenance and repair workers.

- Assembly line workers are often protected by the hazards at the worksite. Maintenance and repair is conducted when the line is down, and hazard controls need not be in place. They also may test equipment when hazard controls are not in place.
- Safety training and PPE is more critical for maintenance and repair workers (refilling chemical tanks, reloading parts, lockout, electrical hazards, use of tools, etc.).
- The highest degree of hazard exists when a robot is in the "teach" mode because the teacher must be within the operating envelope of the robot to program its movements within very close tolerance parameters.

Controls

Companies should conduct ongoing risk analyses to identify hazards in current, new, or redesigned processes and to design effective safety procedures and training programs. JAs must be conducted each time a new machine is installed, whenever there is a process change, whenever a process is automated or removed from automation, or whenever an accident has occurred.

Materials Handling and Storage

Materials handling consists of the movement of any equipment, supplies, bulk containers, raw or finished products, or any other items that must be physically moved from one location to another. This movement is usually tracked from the delivery to storage, to user, or to disposal, or from delivery to storage, to assembly line or manufacture, to final product, to packaging, to storage, or to shipping.

Injuries due to handling of materials accounts for 40 percent of all recorded absences from work, mainly from back injuries, cuts, or strains and sprains. To reduce the number of injuries caused by materials handling, companies should conduct a JA to determine the places and points at which materials are handled, the processes and procedures used to move the materials, the risks associated with those procedures, and the controls required to eliminate or reduce risks.

Using design, some hazards can be controlled or eliminated through automation, use of lifting equipment, or smaller and lighter packaging. Where necessary, administrative controls can be implemented.

The physical hazards of materials handling and storage are:

- worker injury from lifting and handling materials
- spillage or damage of material
- fires or explosions from incompatible storage.

Injuries

Injuries often occur while manually handling materials. Common actions include:

- lifting objects
- carrying objects
- setting down objects
- pushing or pulling objects.

These injuries include:

- strains and sprains from improperly lifting loads or from carrying loads that are either too large or too heavy
- fractures and bruises caused by being struck by materials or by being caught in pinch points
- cuts and bruises caused by falling materials that have been improperly stored or by incorrectly cutting ties or other securing devices (OSHA, 1996c).

Controls for Manual Handling of Materials

Although physical differences make it impractical to establish safe lifting limits for all workers, some general principles can be applied. NIOSH has established an equation to calculate action limits (ALs) and maximum permissible lifts (MPLs) for manual lifting based on vertical and horizontal lifts and the distance traveled. Training workers to lift and handle materials safely is considered an administrative control. Supervision and enforcement is required.

The proper use of PPE, such as leather gloves, safety glasses, hard hats, and steel-toed safety shoes, is a must. However, the use of back belts for lifting has not been proven effective and should only be used if prescribed by a physician. Training must be provided for the usage of accessories for the manual moving and lifting of materials. These include hand-tools (hooks, crowbars, rollers), jacks, dollies, pallet jacks, and hand trucks.

Administrative controls must be enforced. Supervisors are responsible for ensuring that a sufficient numbers of workers are placed on a materials-handling job for two-person lifts and available for assistance as needed. The three main factors to consider when determining the safety of manual moving a particular load are:

- task repetition (How often must the worker move, lift, bend, strain?)
- load location (Where is the load in relation to the worker?)
- load weight (How heavy is each lift?).

Damage to Equipment

Temporary and permanent storage of materials should be neat and orderly to eliminate hazards, make materials easily accessible, and conserve space. Physical hazards that can damage equipment or material include:

- falling, collapsing, or crushing due to improperly stacked or stored material

- fire or explosion due to proximity of incompatible materials or static electricity around metal drums or flammable liquids
- corrosion, leakage, and reactions due to spillage of hazardous liquid chemicals.

Supervisors of shipping and receiving areas must be aware of DOT regulations and labels. Hazardous items will be labeled with DOT-diamond labels providing the hazard category (such as flammable, corrosive, etc.).

Workers in shipping and receiving must be trained in the proper use and handling of such common items as dock boards (truck loading ramps), machines and tools, steel and plastic strapping, burlap and sacking, barrels, kegs, drums, and boxes and cartons. The company must provide safe storage facilities with racks, bins, dividers, cylinder racks, and shelving to accommodate items to be stored.

OSHA recommends using a formal training program to reduce materials handling hazards. Instructors should be well versed in matters that pertain to safety engineering and materials handling and storing. The content of the training should emphasize those factors that will contribute to reducing workplace hazards, including the following:

- alerting the employee to the dangers of lifting without proper training
- showing the employee how to avoid unnecessary physical stress and strain
- teaching workers to become aware of what they can comfortably handle without undue strain
- instructing workers on the proper use of equipment
- teaching workers to recognize potential hazards and how to prevent or correct them.

Because of the high incidence of back injuries, safe lifting techniques for manual lifting should be demonstrated and practiced at the worksite by supervisors as well as by employees. A training program to teach proper lifting techniques should cover the following topics:

- awareness of the health risks to improper lifting, citing organizational case histories
- knowledge of the basic anatomy of the spine, the muscles, and the joints of the trunk and the contributions of intraabdominal pressure while lifting
- awareness of individual body strengths and weaknesses (determining one's own lifting capacity)

- recognition of the physical factors that might contribute to an accident and how to avoid the unexpected
- use of safe lifting postures and timing for smooth, easy lifting and the ability to minimize the load-moment effects
- use of handling aids, such as stages, platforms, steps, trestles, shoulder pads, handles, hand trucks, dollies, and wheels
- knowledge of body responses—warning signals—to be aware of when lifting.

A campaign using posters to draw attention to the need to do something about potential accidents, including lifting and back injuries, is one way to increase awareness of safe work practices and techniques. The plant medical staff and a team of instructors should conduct regular tours of the site to look for potential hazards and allow input from workers (OSHA, 1996a).

Ropes, Chains, Slings, and Rigging

One method of reducing manual–materials-handling injuries is to provide a mechanical means of moving material. Although mechanical lifting aids workers in avoiding manual-lifting injuries, mechanical lifting equipment has

Figure 15.1. Warning sign on a mast swing in a warehouse.

its own physical hazards. Mechanical lifting is normally used to move materials too heavy, bulky, or dangerous to be manually lifted or moved.

OSHA's Rigging Equipment for Material Handling Standard, 29 CFR 1926.251, provides the basic and special safety precautions for the various types of rigging operations.

- The company must have a written policy on how rigging equipment is used and maintained and who is authorized and responsible for the operations.
- Workers must be trained in the special safety precautions required when setting up, using, and storing various ropes, chains, and slings.
- Rigging equipment needs to be weight tested or proof tested, certified, and documented for its intended use. Equipment must be inspected prior to each use.

Types of Rigging

Rigging equipment uses fiber ropes, wire ropes, or chains. Each has a specific use based on their safe working load, durability, and purpose.

Fiber Ropes

- Natural or synthetic fiber ropes are durable but tend to deteriorate when in contact with acids, caustics, and high temperatures.
- Nylon, polyester, and polyolefin ropes are more resistant to corrosive materials, wear and tear, and temperature extremes, although they will melt in a fire.
- Manila ropes are the most widely used natural fiber rope because of superior breaking strength, consistency, and elasticity.
- The main advantages of fiber ropes are their price and their ability to form or bend around angles of the object being lifted.
- Synthetic fiber ropes can be engineered for the purpose, are more resistant to shock loading, and cost more than natural fiber ropes.
- Kevlar fiber rope has superior breaking strength but is nearly four times the cost of natural fiber ropes.

Wire ropes

- Wire rope is a twisted bundle of cold-drawn steel wires.
- It is composed of wires, strands, and a core.
- It is widely used for its greater strength, durability, and predictability of stretch when placed under heavy stresses.
- Wire rope can corrode, kink, and generate sharp, protruding wires.

Supervisors and workers must know the capabilities, hazards, and restrictions of the rope they are using to ensure safety. Dynamic effects are greater on ropes with little stretch and can cause them to break, thus endangering workers. Ropes must be regularly inspected before wear or damage and the results documented. Workers should be trained to handle ropes with care and store them properly, away from harmful substances.

Chains

Steel and alloys (stainless steel, monel, bronze, and other metals) are commonly used for lifting slings made of chain. Each type of chain has a rated capacity or working load limit. Chains for slings will have permanent labels with the size, grade, rated capacity, and sling manufacturer.

Slings

In rigging, various types of ropes, synthetic webbing, and chain are used to fabricate slings. These slings are connected to a hook on a crane or lifting device to move the material. Slings are designed and rated for the types of loads they are to move.

Example: A painted, irregularly shaped object of under one hundred pounds may be moved with a synthetic web-type sling to avoid damage to the paint and to better conform to the object. A heavy pallet would be moved with a wire rope sling attached to the four corners of the square pallet.

The dominant characteristics of a sling are determined by the components of that sling. The strengths and weaknesses of a wire rope sling, for example, are essentially the same as the strengths and weaknesses of the wire rope of which it is made. Fiber rope slings are used for loads that might be damaged by contact with metal, while wire rope and chain slings provide extra strength and durability.

The safety of a sling's assembly depends on the material used, its strength for the load handled, the method of attaching chain to fittings, and proper inspection and maintenance. Alloy steel became the standard material for chain slings because it has high resistance to abrasion and is practically immune to failure because the metal is cold worked. Chain slings are the best choice for lifting very hot materials.

The difference between the usage of synthetic web slings and metal mesh slings is that synthetic web slings are useful for lifting loads that need their surfaces protected while metal mesh slings can safely handle sharp-edged materials, concrete, and high-temperature materials.

Slings are in the physical hazard category of "failures." The hazard from these items is in their failure to perform as designed due to overloading, damage, or excessive wear. Damage to equipment or the material being moved and injury to workers can result from unsafe work practices or improper use of slings. The potential physical energy in a suspended load can be tremendous.

Cautions to Personnel

- Ensure that all portions of the human body are kept away from the area between the sling and the load and between the sling and the crane or hoist hook.
- Ensure that personnel never stand in line with or next to the legs of a sling that is under tension.
- Ensure that personnel do not stand or pass under a suspended load.
- Ensure that personnel do not ride the sling or the load unless the load is specifically designed and tested for carrying personnel.
- Do not use synthetic rope slings as bridles on suspended personnel platforms.
- Do not inspect a wire rope sling by passing bare hands over the wire rope body. Broken wires, if present, may puncture the hands (OSHA, 1996b sling safety).

Summary

Many tasks and hazards must be taken into consideration when working in a retail, warehouse, or service environment. The main workplace injury in these industries involved CTDs usually related to repetitive-type motions involved with lifting, bending, or carrying loads. CTDs can be reduced through the company's commitment to safety. From day one on the job, employees can be introduced to the company's safety culture and programs.

Review Questions

1. List common physical hazards associated with service, retail, and warehouse industries.
2. How can companies in retail and distribution manage their risks and losses?
3. What topics should be included in an employee safety orientation program?
4. Transportation safety should include which areas of travel?
5. Materials handling poses many risks to the worker. Give several examples.

16

Workplace Violence

---

3 Dead in Albuquerque Office Rampage
July 12, 2010 | By Michael Haederle, Los Angeles Times
A gunman targeting his live-in girlfriend opened fire at a fiberoptics manufactur-
ing plant Monday, killing two people and wounding four others before turning
the weapon on himself, police say. The gunman was identified by police as Robert
Reza, a former employee of Emcore Corp., where hundreds of workers fled after
the shooting broke out shortly before 9:30 a.m. "We believe it is a workplace
domestic violence situation," Albuquerque Police Chief Ray Schultz said, adding
that the girlfriend, who had told coworkers that she feared for her safety, was
among those wounded.

UNFORTUNATELY HEADLINES LIKE THIS are all too common; it seems every
month or two one of these sensational reports make the leading story in
the news media. Workplace violence has become a top concern for safety and
health professionals throughout the country.

What Is Workplace Violence?

According to the Centers for Disease Control (1996), defining workplace
violence has generated considerable discussion. Some would include in the
definition any language or actions that make one person uncomfortable in
the workplace, others would include threats and harassment, and all would

include any bodily injury inflicted by one person on another. Thus the spectrum of workplace violence ranges from offensive language to homicide, and a reasonable working definition of workplace violence is as follows: *violent acts, including physical assaults and threats of assault, directed toward persons at work or on duty.* Most studies to date have focused primarily on physical injuries because they are clearly defined and easily measured.

The circumstances of workplace violence also vary and may include robbery-associated violence; violence by disgruntled clients, customers, patients, inmates, and so on; violence by coworkers, employees, or employers; and domestic violence that finds its way into the workplace. These circumstances all appear to be related to the level of violence in communities and in society in general. Thus the question arises: why study workplace violence separately from the larger universe of all violence? Several reasons exist for focusing specifically on workplace violence. Violence is a substantial contributor to death and injury on the job. NIOSH data indicate that homicide has become the second-leading cause of occupational injury death, exceeded only by motor-vehicle–related deaths. Estimates of nonfatal workplace assaults vary dramatically, but a reasonable estimate from the National Crime Victimization Survey is that approximately 1 million people are assaulted while at work or on duty each year; this figure represents 15 percent of the acts of violence experienced by U.S. residents aged twelve or older.

The circumstances of workplace violence differ significantly from those of all homicides. For example, 75 percent of all workplace homicides in 1993 were robbery related; but in the general population, only 9 percent of homicides were robbery related, and only 19 percent were committed in conjunction with any kind of felony (robbery, rape, arson, etc.). Furthermore, 47 percent of all murder victims in 1993 were related to or acquainted with their assailants, whereas the majority of workplace homicides (because they are robbery related) are believed to occur among persons not known to one another. Only 17 percent of female victims of workplace homicides were killed by a spouse or former spouse, whereas 29 percent of the female homicide victims in the general population were killed by a husband, ex-husband, boyfriend, or ex-boyfriend.

Workplace violence is not distributed randomly across all workplaces but is clustered in particular occupational settings. Out of 421 workplace shootings recorded in 2008 (8 percent of total fatal injuries), 99 (24 percent) occurred in service and retail trade. Homicide is the leading cause of death in these industries, as well as in finance, insurance, and real estate. Eighty-five percent of nonfatal assaults in the workplace occur in service and retail trade industries. Workplace shootings in manufacturing were less common, with seventeen shootings reported in 2008 (BLS, 2010). As the U.S. economy continues to shift toward the service sectors, fatal and nonfatal workplace violence will be an increasingly important occupational safety and health issue.

Controls

Because the retail and service trades are the most highly affected by workplace violence, we will focus on the controls needed in these work environments. The risk of workplace violence is associated with specific workplace factors, such as dealing with the public, the exchange of money, and the delivery of services or goods. Consequently, great potential exists for workplace-specific prevention efforts, such as bullet-resistant barriers and enclosures in taxicabs, convenience stores, gas stations, emergency departments, and other areas where workers come in direct contact with the public; locked drop safes and other cash-handling procedures in retail establishments; and threat assessment policies in all types of workplaces. Additional workplace controls are discussed throughout the chapter.

Engineering and Environmental Controls

The CDC (1996) recommends the following:

Physical separation of workers from customers, clients, and the general public through the use of bullet-resistant barriers or enclosures has been proposed for retail settings such as gas stations and convenience stores, hospital emergency departments, and social service agency claims areas. The height and depth of counters (with or without bullet-resistant barriers) are also important considerations in protecting workers, since they introduce physical distance between workers and potential attackers. Consideration must nonetheless be given to the continued ease of conducting business; a safety device that increases frustration for workers or for customers, clients, or patients may be self-defeating.

Visibility and lighting are also important environmental design considerations. Making high-risk areas visible to more people and installing good external lighting should decrease the risk of workplace assaults.

Access to and egress from the workplaces are also important areas to assess. The number of entrances and exits, the ease with which nonemployees can gain access to work areas because doors are unlocked, and the number of areas where potential attackers can hide are issues that should be addressed. This issue has implications for the design of buildings and parking areas, landscaping, and the placement of garbage areas, outdoor refrigeration areas, and other storage facilities that workers must use during a work shift.

Numerous security devices may reduce the risk for assaults against workers and facilitate the identification and apprehension of perpetrators. These include closed-circuit cameras, alarms, two-way mirrors, card-key access systems, panic-bar doors locked from the outside only, and trouble lights or geographic locating devices in taxicabs and other mobile workplaces.

Administrative Controls

Staffing plans and work practices (such as escorting patients and prohibiting unsupervised movement within and between clinic areas) are included in the California Occupational Safety and Health Administration *Guidelines for the Security and Safety of Health Care and Community Service Workers*. Increasing the number of staff on duty may also be appropriate in any number of service and retail settings. The use of security guards or receptionists to screen persons entering the workplace and controlling access to actual work areas has also been suggested by security experts.

Work practices and staffing patterns during the opening and closing of establishments and during money drops and pickups should be carefully reviewed for the increased risk of assault they pose to workers. These practices include having workers take out garbage, dispose of grease, store food or other items in external storage areas, and transport or store money.

Policies and procedures for assessing and reporting threats allow employers to track and assess threats and violent incidents in the workplace. Such policies clearly indicate a zero tolerance of workplace violence and provide mechanisms by which incidents can be reported and handled. In addition, such information allows employers to assess whether prevention strategies are appropriate and effective. These policies should also include guidance on recognizing the potential for violence, methods for defusing or de-escalating potentially violent situations, and instruction about the use of security devices and protective equipment. Procedures for obtaining medical care and psychological support following violent incidents should also be addressed. Training and education efforts are clearly needed to accompany such policies.

Training employees in nonviolent response and conflict resolution has been suggested to reduce the risk that volatile situations will escalate to physical violence. Also critical is training that addresses hazards associated with specific tasks or worksites and relevant prevention strategies. Training should not be regarded as the sole prevention strategy but as a component in a comprehensive approach to reducing workplace violence. To increase vigilance and compliance with stated violence prevention policies, training should emphasize the appropriate use and maintenance of protective equipment, adherence to administrative controls, and increased knowledge and awareness of the risk of workplace violence.

Personal Protective Equipment (PPE)

PPE, such as body armor, has been used effectively by public safety personnel to mitigate the effects of workplace violence. For example, the lives of more than 1,800 police officers have been saved by Kevlar vests.

Long-term efforts to reduce the level of violence in U.S. society must address a variety of social issues, such as education, poverty, and environmental justice. However, short-term efforts must address the pervasive nature of violence in our society and the need to protect workers. We cannot wait to address workplace violence as a social issue alone but must take immediate action to address it as a serious occupational safety issue.

Implementing a Violence-in-the-Workplace Prevention Program

The first priority in developing a workplace violence prevention policy is to establish a system for documenting violent incidents in the workplace. Such data are essential for assessing the nature and magnitude of violence in a given workplace and quantifying risk. These data can be used to assess the need for action to reduce or mitigate the risks for workplace violence and implement a reasonable intervention strategy. An existing intervention strategy may be identified within an industry or in similar industries, or new and unique strategies may be needed to address the risks in a given workplace or setting. Implementation of the reporting system, a workplace violence prevention policy, and specific prevention strategies should be publicized companywide, and appropriate training sessions should be scheduled. The demonstrated commitment of management is crucial to the success of the program. The success and appropriateness of intervention strategies can be monitored and adjusted with continued data collection.

A written workplace violence policy should clearly indicate a zero tolerance of violence at work, whether the violence originates inside or outside the workplace. Just as workplaces have developed mechanisms for reporting and dealing with sexual harassment, they must also develop threat assessment teams to which threats and violent incidents can be reported. These teams should include representatives from human resources, security, employee assistance, unions, workers, management, and perhaps legal and public relations departments. The charge to this team is to assess threats of violence (e.g., to determine how specific a threat is, whether the person threatening the worker has the means for carrying out the threat, etc.) and to determine what steps are necessary to prevent the threat from being carried out. This team should also be charged with periodic reviews of violent incidents to identify ways in which similar incidents can be prevented in the future. Note that when violence or the threat of violence occurs among coworkers, firing the perpetrator may or may not be the most appropriate way to reduce the risk for additional or future violence. The employer may want to retain some control over the perpetrator and require or provide counseling or other care, if appropriate. The violence prevention policy should explicitly state the consequences of making threats or committing acts of violence in the workplace.

A comprehensive workplace violence prevention policy and program should also include procedures and responsibilities to be taken in the event of a violent incident in the workplace. This policy should explicitly state how the response team is to be assembled and who is responsible for immediate care of the victim(s); reestablishing work areas and processes; and organizing and carrying out stress-debriefing sessions with victims, their coworkers, and perhaps the families of victims and coworkers. Employee assistance programs, human resource professionals, and local mental health and emergency service personnel can offer assistance in developing these strategies.

Responding to an Immediate Threat of Workplace Violence

For a situation that poses an immediate threat of workplace violence, all legal, human resource, employee assistance, community mental health, and law enforcement resources should be used to develop a response. The risk of injury to all workers should be minimized. If a threat has been made that refers to particular times and places, or if the potential offender is knowledgeable about workplace procedures and timeframes, patterns may need to be shifted. For example, a person who has leveled a threat against a worker may indicate, "I know where you park and what time you get off work!" In such a case, it may be advisable to change or even stagger departure times and implement a buddy system or an escort by security guard for leaving the building and getting to parking areas. The threat should not be ignored in the hope that it will resolve itself or out of fear of triggering an outburst from the person who has lodged the threat. If someone poses a danger to himself or others, appropriate authorities should be notified and action should be taken.

Dealing with the Consequences of Workplace Violence

Much discussion has also centered around the role of stress in workplace violence. The most important thing to remember is that stress can be both a cause and an effect of workplace violence. That is, high levels of stress may lead to violence in the workplace, but a violent incident in the workplace will most certainly lead to stress, perhaps even to posttraumatic stress disorder. The data from the National Crime Victimization Survey (Bachman, 1994) present compelling evidence (more than a million workdays lost as a result of workplace assaults each year) for the need to be aware of the impact of workplace violence. Employers should therefore be sensitive to the effects of workplace violence and provide an environment that promotes open communication; they should also have in place an established procedure for reporting and responding to violence. Appropriate referrals to employee assistance programs or other local mental health services may be appropriate for stress-debriefing sessions after critical incidents (CDC, 1996).

Summary

Because violence in the workplace has become so pervasive, it has also become an issue for the safety professional. Violence can come in the form of a threat and escalate to mass murders. Employees must be trained in recognizing the potential for dangerous situations and persons in the workplace.

Review Questions

1. What is workplace violence?
2. Name the types of workplaces that are more susceptible to violence.
3. List some controls that may help prevent violence in the workplace.
4. What is the importance of a violent-acts–reporting system?
5. What resources need to be available to employees?

17

Safety Evaluations
and Inspection Processes

PHYSICAL HAZARDS ARE IDENTIFIED during the job analysis, the risks assessed, and the safety requirements or controls determined and put into effect. The controls may include designs to eliminate or mitigate hazards but may also require administrative controls and/or personal protective equipment. Once the controls are applied, a routine program of inspections and reevaluation is necessary to ensure:

- safety controls remain in place and are effective
- changes have not been made to those controls or processes
- safety requirements are being complied with and enforced
- there is documentation of inspections in case of an accident or to support the accident investigation.

Engineering design to eliminate or reduce physical hazards still requires worker influence and interface to remain safe. Use of administrative controls and PPE take frequent supervision to ensure compliance. The only way to verify a workplace remains safe is to inspect.

Compliance

Safety compliance involves conforming to local, state, and federal safety regulations, as well as any company policies. Federal regulations are usually minimum standards written with a broad applicability—because there are thousands of situations, companies, and industries to which they must apply. State and local regulations add emphasis or more rules to the minimum standards; they can never reduce the requirements of a federal standard. State and local regulations may also add or clarify regulations to be site specific or locality specific. For example, a local regulation may address requiring annual testing for radon when determining indoor air quality if radon in the bedrock is of higher risk in that area.

Inspections must include evaluating a process, workplace, or procedure according to the required regulations. The safety professional must determine which regulations pertain and be able to cite that regulation as part of the inspection results.

Federal OSHA and State OSH Inspections

Federal OSHA does not have the staff to inspect all 5 million–plus workplaces in the United States. Federal OSHA is tasked with inspecting federally owned and operated government properties, military bases, and institutions. Twenty-six states have their own OSH administrations. These are approved by federal OSHA and monitored. Where there are no state OSH administrations, federal OSH will monitor high-risk industries via regional offices and conduct investigations of serious or fatal accidents. They also monitor industries with high numbers of lost workdays or accidents.

The key to federal or state OSHA inspections is to know the applicable regulations and conduct a regular program of inspections to ensure management is aware of hazards and safety issues.

The OSHA Inspection Process

Companies are notified that they will be having a compliance inspection. Where security is an issue (government contractor work, for example), security information will be sent ahead so the inspector can gain access.

- The inspection starts with an opening conference or brief.
- The inspector is introduced to management and supervisors.
- The inspection process, scope, and standards to be used are explained.
- Company and union personnel to accompany the inspector will be designated.

The compliance inspector and escorts will start the walk-around inspection by doing the following:

- observing workers but trying not to interrupt production
- possibly taking photos for reference
- documenting findings
- possibly talking to workers (employee safety and health rights are protected)
- checking documentation (JA!), records, and investigations
- immediately correcting dangerous situations on the spot.

The inspection may not cover the entire plant if investigating for a specific incident, such as a fatality.

A closing conference or debrief is held at the end of the inspection to notify management of their findings.

- Violations are discussed.
- Appeal rights are discussed.
- The penalties are not discussed—those are determined by the area director after report review.

Citations and Penalties

After the walkthrough, a complete list of violations citing the applicable regulations is generated. This report is reviewed by the OSHA area director to see if any penalties are to be assigned. Penalties can range from a few thousand dollars to seventy thousand dollars.

There are four types of OSHA violations:

1. other than serious violation—has a direct relationship to job safety and health but low risk (will not cause death or serious injury)
2. serious violation—substantial probability that death or serious physical harm would occur and where the employer knew or should have known of the hazard (*must* propose a penalty)
3. willful violation—violation where the employer intentionally and knowingly commits a violation or is aware that a hazardous condition exists and has made no reasonable effort to eliminate it (A proposed penalty is always five thousand dollars or more and may include legal issues.)
4. repeated violation—violation of a standard, regulation, rule, or order where, upon reinspection, a substantially similar or the same violation was found.

Penalties are determined based on four factors:

1. gravity of the violation
2. size of the business
3. good faith of the employer
4. employer's history of previous violations.

Employers can contest the violations and penalties and may have the dollar amounts lowered. There is an appeal process to discuss and work out resolving a citation, modifying the requirement to fix the violation, extending the time to fix the violation, and taking the case as high as the Department of Labor for review. Where it is a state OSH violation, the states have a similar review and appeal process.

Employee Safety Complaints

When an OSHA inspection is the result of a safety complaint by an employee, OSHA responds with the findings back to that complainant. If a complaint is made and OSHA inspects but does not issue a citation or issues a lesser citation and the employee thinks this is wrong or disagrees, the worker has ten days to contest the decision. Within fifteen days of his or her employer receiving the citation, the employee may submit a written objection to OSHA. In some cases, the employee's union may provide support and assistance with the appeal process. This is a right, and the employees must be made aware of this as part of safety training.

Part of the plant or company's written safety program should include information on how to deal with OSHA inspections, employee safety complaints, and appeals. The company should have its own safety inspection program to review the same compliance areas as OSHA to avoid surprises.

OSHA Employee Rights

According to OSHA, employee rights include the following:

- The employee has the right to notify his or her employer or OSHA about workplace hazards. The employee may ask OSHA to keep his or her name confidential.
- The employee has the right to request an OSHA inspection if he or she believes that there are unsafe and unhealthful conditions in the

workplace. The employee or his or her representative may participate in that inspection.

- The employee has the ability to file a complaint with OSHA within thirty days of retaliation or discrimination by his or her employer for making safety and health complaints or for exercising his or her rights under the OSH Act.
- The employee has a right to see OSHA citations issued to his or her employer. The employer must post the citations at or near the place of the alleged violation.
- The employer must correct workplace hazards by the date indicated on the citation and must certify that these hazards have been reduced or eliminated.
- The employee has the right to copies of his or her medical records or records of his or her exposure to toxic and harmful substances or conditions.
- The employer must post this notice in the workplace.
- The employee must comply with all OSH standards issued under the OSH Act that apply to his or her own actions and conduct on the job.

According to OSHA, employers must furnish employees a place of employment free from recognized hazards and comply with the OSH standards issued under the OSH Act.

Safety and Health Training

A significant element in control of physical hazards is training of the workers to recognize the hazards, understand how the controls are applied, and how to comply with the safety requirements. OSHA requires all workers be trained on the hazards of their job. Safety training is driven by regulations.

At the beginning of this book, it is stated that controls involve continuous improvement and redesign to reduce the number of errors until operations are as error-proof as possible. It is also said that safety by design seeks to reduce human decision making as part of the procedure to reduce errors. The process to reduce errors, or reduce decision-based errors, depends partly on safety education and training.

- Education is knowledge based (why we use machine guards and how they are designed to work).
- Training is skills based (how the machine guard attaches and is used).

Training is focused mainly on behavior change, showing workers how to do something properly and to apply their knowledge on the job. Adult learners have special needs and requirements that trainers must recognize for the programs to be effective. Ensure that training is tailored to the employee's level of knowledge (can they read?), safety experience, previous training, and the objective of the training. To design an effective training program, the safety professional must assess workers' needs, analyze learners' characteristics, develop specific objectives, develop materials and schedules, and design testing and evaluation methods. Training begins with new employee orientation, on-the-job training, job instruction, group methods, and individual methods. Providing a written policies and procedures manual is one way of meeting new employee training needs and of conforming to regulatory standards.

Identifying Training Needs

Training needs can be identified during the JA. Safety professionals can use this time to talk to the employees to get a sense of safety knowledge and worker skill level, and this will help determine where there is a need for training. Additional training needs may be pointed out:

- when new equipment is received, when there are new work procedures put into place, or when work procedures are changed
- after making safety improvements (Perhaps certain precautions no longer apply due to design changes!)
- when workers change jobs or worksites, when new workers are hired, or after an accident has occurred (review)
- when safety regulations or rules change—train to the regulations.

Safety training products are available from many sources and can be easily tailored to the company's needs.

Strict records must be kept of all safety training and should include worker name, date completed, who administered the training, and the type of safety training/topics covered. This may be critical in the case of an accident investigation. Indirect training is used to supplement safety training programs. Safety training should be focused on specific tasks and procedures; additional references should include:

- material safety data sheets
- flowcharts, checklists, and diagrams
- troubleshooting guides

- reference manuals
- helpdesks or hotlines
- reward systems and positive reinforcement.

Safety training is a significant element of hazard control; it supplements and compliments design changes and engineering controls, reinforces administrative controls, and is important in the proper care and use of PPE. Engineering controls and control through design may serve to decrease overall training requirements as hazards are decreased.

Citing Safety Standards and Regulations

Safety professionals must know and understand—and know where to find—the standards, laws, and regulations that cover the compliance and safety hazards they identify. While the professional opinion may be necessary at times, and an interpretation or explanation is desired, nothing can replace citing the exact passage and providing the source.

Most of the safety standards followed by industry are contained in the OSHA regulations. Even state standards must meet or exceed federal regulations or established federal standards. An established federal standard means any operative standard established by any agency of the United States and in effect on April 28, 1971, or contained in any act of Congress in force on the date of enactment of the Williams-Steiger OSH Act.

Federal standards are laws found in the *Code of Federal Regulations*. For safety and occupational health general industry standards, the *Code of Federal Regulations* title 29, part 1910, is used. Part 1910 is further divided into subparts, sections, paragraphs, and subparagraphs. The 29 CFR 1910 has subparts A through Z, each addressing a separate set of standards. For example, 29 CFR 1910, subpart F, covers powered platforms, man lifts, and vehicle-mounted work platform safety regulations. Subpart L addresses fire protection requirements. When citing an OSHA standard in an inspection or survey, we need to define the title, part, and paragraph well enough that the exact section can be located. The OSHA regulations are contained in published paperback books—dozens of them in a set—or they can be accessed and searched on the Internet at www.osha.gov.

Example: During a hazard evaluation, it is found that the workers are not using eye protection on the job. One worker said he did not think he needed safety glasses, and management was not enforcing it. The safety professional writes up the discrepancy and cites the OSHA standard:

29 CFR 1910.133(a)(1) "The employer shall ensure that each affected employee uses appropriate eye or face protection when exposed to eye or face hazards from flying particles, molten metal, liquid chemicals, acids or caustic liquids, chemical gases or vapors, or potentially injurious light radiation."

The *Code of Federal Regulations* title 29, part 1910, paragraph 133, subparagraph (a) (1) was cited.

In some cases, the design and performance of a particular piece of safety equipment is referred to in an OSHA standard. OSHA standard 1910.133(b)(1) says, "Protective eye and face devices purchased after July 5, 1994 shall comply with ANSI [American National Standards Institute] Z87.1-1989, 'American National Standard Practice for Occupational and Educational Eye and Face Protection,' which is incorporated by reference as specified in Sec. 1910.6." The ANSI standard will have to be referenced for specifics.

Example: An employee wears sunglasses instead of safety glasses and demands that you show him or her, in writing, where it says he *can't* wear sunglasses instead of the safety glasses.

ANSI and other organizations that develop standards and regulations are cited in a similar manner. ANSI/ASSE Z-87.1-2003 is the ANSI and American Society of Safety Engineers design and performance standard for safety glasses. The "Z" is the section, numbered in sequence with a subpart ".1," and the year it was issued is "2003." All must be cited if used as a reference.

The American Society for Testing and Materials (ASTM) also issues standards on the construction, specifications, and test methods for equipment, including safety devices. For example, ASTM C730-98(2003) is the Standard Test Method for the Knoop Indentation Hardness of Glass to test the lenses used to make safety glasses.

In order to ensure compliance by company officials, safety professionals must be familiar with the laws, regulations, and standards that apply to their companies and be prepared to cite those references.

Summary

For the safety professional, it is important to know where to find federal regulations. Safety regulations and standards are the backbone and the main resources for all safety-related work and enforcement at the worksite. Without knowing the regulations or where to find them, a safety professional would not have a leg to stand on.

Review Questions

1. What is the key to federal or state OSHA inspections?
2. Explain the proceedings of the OSHA inspection process.
3. When does safety training need to occur?
4. Define the four types of OSHA violations.
5. What are the types of OSHA penalties?

References and Recommended Reading

American National Standards Institute (ANSI) and American Society of Mechanical Engineers (ASME). (2007). A17.2-2007, *Guide for Inspection of Elevators, Escalators, and Moving Walks.* New York: ASME International.

Associated Press. (2010, January 25). "N.Y. Janitor Crushed to Death in Trash Compactor." *Fox News Network, LLC.* Retrieved September 5, 2010, from http://www.foxnews.com/us/2010/07/25/maintenance-man-crush-death-ny-trash-compactor

Bachman, Ronet. (1994, January). "Violence against Women: A National Crime Victimization Survey Report." U.S. Department of Justice, Bureau of Justice Statistics.

Bureau of Labor Statistics (BLS). (2003, January 30). *Compensation and Working Conditions; Amputations: A Continuing Workplace Hazard.* United States Department of Labor; Bureau of Labor Statistics. Retrieved September 5, 2010, from http://www.bls.gov/opub/cwc/sh20030114ar01p1.htm

Bureau of Labor Statistics (BLS). (2009, December 17). *Occupational Outlook Handbook, 2010–11 Edition: Welding, Soldering, and Brazing.* Retrieved September 5, 2010, from http://www.bls.gov/oco/ocos226.htm

Bureau of Labor Statistics (BLS). (2010, July 14). *Fact Sheet: Workplace Shootings.* United States Department of Labor. Retrieved September 16, 2010, from http://www.bls.gov/iif/oshwc/cfoi/osar0014.htm

California Occupational Safety and Health Administration. (1998, March 10). *Guidelines for the Security and Safety of Health Care and Community Service Workers,* http://www.dir.ca.gov/dosh/dosh_publications/hcworker.html.

Centers for Disease Control (CDC). (1996, July). *Current Intelligence Bulletin 57: Violence in the Workplace.* United States Department of Labor. Retrieved September 16, 2010, from http://www.cdc.gov/niosh/violpurp.html

Centers for Disease Control (CDC). (2006, May 10). *Radiation Emergencies: Radiation Dictionary.* Retrieved September 7, 2010, from http://www.bt.cdc.gov/radiation/glossary.asp

Centers for Disease Control (CDC). (2010, May 3). *Workplace Safety and Health Topics: Ergonomics and Musckuloskeletal Disorders.* Retrieved September 5, 2010, from http://www.cdc.gov/niosh/topics/ergonomics

Department of Health and Human Services (DHHS) and National Institute of Occupational Health and Safety (NIOSH). (1988, April). *DHHS (NIOSH) Publication No. 88-110: NIOSH Criteria Documents, Criteria for a Recommended Standard: Welding, Brazing, and Thermal Cutting.* Center for Disease Control. Retrieved September 7, 2010, from http://www.cdc.gov/niosh/88-110.html

Haederle, M. (2010, July 12). "3 Dead in Albuqerque Office Rampage." *The Los Angeles Times.*

House Hearing Institute (HEI). (2008). *Common Focus of Hearing Loss.* Retrieved December 12, 2010, from http://www.hei.org./education/health/loss.htm

Massachusetts Department of Public Health, Occupational Health Surveillance Program. (1997). *Fatality Investigation Report 2:2.* Boston: Massachusetts Department of Public Health.

National Fire Protection Agency (NFPA). (2009a). *NFPA 70E: Standard for Electrical Safety in the Workplace.* New York: NFPA.

National Fire Protection Association (NFPA). (2009b). *NFPA 101: Life Safety Code.* New York: NFPA.

National Institute for Occupational Safety and Health (NIOSH). (1986). *Preventing Occupational Fatalities in Confined Spaces—NIOSH #86-110.* Centers for Disease Control. Retrieved December 22, 2010, from http://www.cdc.gov/NIOSH/86110v2.html

National Institute for Occupational Safety and Health (NIOSH). (1995). *Fatality Assessment and Control Evaluation (FACE) Report No. 95-12: Laborer Fatally Injured while Cleaning Concrete Mixer—Tennessee.* Morgantown, WV: U.S. Department of Health and Human Services, Public Health Service, Centers for Disease Control and Prevention, NIOSH.

National Institute of Occupational Safety and Health (NIOSH). (1996). *NIOSH Publication No. 96-110: Preventing Occupational Hearing Loss—A Practical Guide 1996.* Centers for Disease Control. Retrieved April 24, 2008, from http://www.cdc.gov/niosh/docs/96-110/emerging.html

National Institute for Occupational Safety and Health (NIOSH). (1999). *DHHS (NIOSH) Publication No. 99-110; NIOSH Alert: Preventing Worker Deaths from Uncontrolled Release of Electrical, Mechanical, and Other Types of Hazardous Energy.* Cincinnati: NIOSH.

National Institute of Occupational Safety and Health (NIOSH). (2001, June). *NIOSH Publication No. 2001-109: Preventing Injuries and Deaths of Workers.* Centers for Disease Control. Retrieved September 6, 2010, from http://www.cdc.gov/niosh/docs/2001-109/default.html

National Institute of Occupational Safety and Health (NIOSH). (2007, August). *NIOSH Publication No. 2007-122: Simple Solutions: Ergonomics for Construction*

Workers. Retrieved September, from http://www.cdc.gov/niosh/docs/2007-122/default.html

National Institute of Occupational Health and Safety (NIOSH) and Occupational Health and Safety Administration (OSHA). (1979, December 4). *DHHS (NIOSH) Publication No. 80-107: Radiofrequency Sealers and Heaters: Potential Health Hazards and Their Prevention.* Retrieved Septemeber 10, 2010, from http://www.cdc.gov/niosh/80107_33.html

Neitzel, R., Seixas, N., Camp, J., Yost, J., and Ren, K. (n.d.). *Occupational Noise Exposures in Five Construction Trades.* University of Washington, Department of Environmental Health. Retrieved April 24, 2008, from http://staff.washington.edu/rneitzel/Occupational_Noise_5trades.pdf

Nelson, D., Nelson, R., Concha-Barrientos, M., and Fingerhut, M. (2005). "The Global Burden of Occupational Noise-Induced Hearing Loss." *American Journal of Industrial Medicine.* Retrieved April 24, 2008, from http://www.who.int/quantifying_ehimpacts/global/6noise.pdf

Occupational Safety and Health Administration (OSHA). (1991, August 5). *OSHA Archive Instruction: STD 01-05-001: Guidelines for Laser Safety and Hazard Assessment.* U.S. Department of Labor. Retrieved September 8, 2010, from http://63.234.227.130/pls/oshaweb/owadisp.show_document?p_id=1705&p_table=DIRECTIVES

Occupational and Health Administration (OSHA). (1995). *Ionizing Radiation.* 29 CFR 1910.1096. U.S. Department of Labor. Retrieved December 12, 2010, from http://www.osha.gov/pls/oshaweb/owadisp.show_document?p_table=STANDARDS&p_id=10098

Occupational Safety and Health Administration (OSHA). (1996a, June 13). *OSHA Hazard Information Bulletin: Asphyxiation Hazard in the Pits: Potential Confined Space Problem.* U.S. Department of Labor. Retrieved September 18, 2010, from http://63.234.227.130/dts/hib/hib_data/hib19960613.html

Occupational and Health Administration (OSHA). (1996b). OSHA 3072: *Sling Pamphlet.* U.S. Department of Labor. http://www.osha.gov/Publications/osha3072.html

Occupational Safety and Health Administration (OSHA), Office of Training and Education. (1996c, May). *Welding Health Hazards: Construction Safety and Health Outreach Program.* U.S. Department of Labor. Retrieved September 7, 2010, from http://63.234.227.130/doc/outreachtraining/htmlfiles/weldhlth.html

Occupational Safety and Health Administration (OSHA). (1999). *OSHA 3157: Guide for Protecting Workers from Woodworking Hazards.* U.S. Department of Labor. Retrieved September 6, 2010, from http://63.234.227.130/Publications/woodworking_hazards/osha3157.html

Occupational Safety and Health Administration (OSHA). (2002). *OSHA 3071: Hearing Conservation.* U.S. Department of Labor. Retrieved September 12, 2010, from http://63.234.227.130/Publications/osha3074.pdf

Occupational Safety and Health Administration (OSHA). (2005, May 4). *Safety and Health Topics: Radiofrequency and Microwave Radiation.* U.S. Department of Labor. Retrieved September 8, 2010, from http://63.234.227.130/SLTC/radiofrequencyradiation/index.html

Occupational Safety and Heath Administration (OSHA). (2007a, March 9). *Safety and Health Topics: Control of Hazardous Energy (Lockout/Tagout).* U.S. Department of Labor. Retrieved September 5, 2010, from http://63.234.227.130/SLTC/controlhazardousenergy/index.html

Occupational Health and Safety Administration (OSHA). (2007b, August 14). *Safety and Health Topics: Confined Spaces.* U.S. Department of Labor. Retrieved September 18, 2010, from http://63.234.227.130/SLTC/confinedspaces/index.html

Occupational Safety and Health Administration (OSHA). (2007c, August 1). *Safety and Health Topics: Metalworking Fluids.* U.S. Department of Labor. Retrieved September 7, 2010, from http://63.234.227.130/SLTC/metalworkingfluids/index.html

Occupational Safety and Health Administration (OSHA). (2007d, July 5). *Safety and Health Topics: Welding, Cutting, and Brazing.* U.S. Department of Labor. Retrieved September 7, 2010, from http://63.234.227.130/SLTC/weldingcuttingbrazing

Occupational Safety and Heath Administration (OSHA). (2008). *OSHA 3341-03N: Guidelines for Shipyards: Ergonomics for the Prevention of Musculoskeletal Disorders.* U.S. Department of Labor. Retrieved September 5, 2010, from http://63.234.227.130/dsg/guidance/shipyard-guidelines.html

Occupational Safety and Heath Administration (OSHA). (2010, October 22). *29 CFR 1910.* GPO Access: Electronic Code of Federal Regulations. Retrieved October 22, 2010, from http://ecfr.gpoaccess.gov/cgi/t/text/text-idx?c=ecfr&tpl=/ecfrbrowse/Title29/29cfr1910_main_02.tpl

U.S. Department of Energy (USDOE). (1994, December). *DOE-STD-5503-94: DOE Standard: EM Health and Safety Plan Guidelines.* Retrieved September 18, 2010, from http://www.hss.doe.gov/nuclearsafety/ns/techstds/standard/est5503/est5503.pdf

Wald, P. H., and Stave, G. M. (2002). *Physical and Biological Hazards of the Workplace.* New York: John Wiley & Sons, Inc.

Index

ADA, 17–19
administrative controls, 2, 32–33, 41, 52, 94, 130
ALARA, 48
annual audiogram, 44
ANSI color-coded signs, 9–10
anthropometry, 54
audiometric testing, 43
automated guided vehicle, 101–103
automated lines, systems, or processes, 118–119

baseline audiogram, 43–44
"big-box" stores, 112
biological hazards, 6
biomechanics, 54
boilers, 74–76
brazing, 96–97
broom test, 76

carpal tunnel syndrome, 54
chains, 125
chemical hazards, 20
chimneys, 13
chronic back pain, 54
Class A fires, 67

Class B fires, 67
Class C fires, 67
Class D fires, 67
cold forming metals, 92–93
cold weather gear, 35
color codes, 9
combustible liquids, 71–73
combustion, 8–19
compliance, 134–135
compressed gas cylinders, 99–100
computerized preventive maintenance (CPM), 15
confined spaces, 28–39
    hot work in, 94
    permit-required spaces, 29–33
construction standards, 79
conveying equipment, 107–109
cooling system, 35
CPR, 64
cranes, 108
crimpers, 89
CTDs, 78
current, 59
cutting, 96
cutting fluid toxicity, 90

design, 4–5
drill presses, 89
driveways, 13

electrical energy, 59
electrical safety, 59–64
electrocution, 60
elevators, 109
engineering controls, 2, 32–33, 41, 48,
    84, 90, 113, 117, 129
environmental controls, 129
ergonomics, 54–58
    elements of, 56
    risk factors, 57
    training, 56
escalators, 109
excavations, 11
exits, 10, 14
extremely low frequency, 49

feeding methods, 23
fire classifications, 67
fire detection, 68
fire doors, 68
fire prevention, 86–87
fire safety, 66–76
firewalls, 68
fixed guards, 62
flammable liquids, 71–73
flammable storage lockers, 69
flexible hoses, 76
flight safety, 116–117
floors, 13
flying metal chips, 90
flying tools, 83
foam system, 70

gates, 23
good housekeeping, 87
ground fault interrupter, 64
grounding, 61
guard design, 23–24
guards, 22

hand-tools, 78–79
haulage equipment, 106–107

hazard, 1
    control of, 2–3
hazard analysis, 5
hazard control, 3–5
hearing conservation program, 42–46
    protection requirement, 45
    recordkeeping for, 46
    training for, 45–46
hoisting equipment, 107–109
hoists, 12
holes, 11–12
hot surfaces, 73
hot work, 93–94
hot working metals, 93

immediately dangerous to life and
    health (IDLH), 30, 35–36
immediate threat, 132
industrial sanitation, 15
inspection, 134–142
interlocks, 22

Job Hazard Analysis (JHA), 5–7, 11, 55,
    61, 67, 74–75, 113, 119, 139

kickbacks, 82, 84

ladders, 12
laser hazards, 49
    controls of, 50
    sources of, 49
lathes, 89
layout, 8–19
light-reflectance value (LRV), 9
liquefied petroleum gas, 103
loading docks, 13
lockout and tagout (LOTO), 24–27,
    75
lockouts, 20–27

machine action, 22
machine guards, 84
machine hazards, 82
machine motions, 21–22
machinery work practice controls, 91
maintenance, 8–19

marine terminals, 117
materials handling, 120–123
metal cutoff saws, 89
metallic guards, 73
metalworking fluids, 91–92
metalworking machinery, 89–95
microwaves, 49
mills, 89

noise, 40–46
    controls, 41–42
    monitoring of, 42
non-ionizing radiation, 48–49
nonmechanical hazards, 22

off-road equipment, 106–107
openings, 10–12
OSHA (Occupational Safety and
    Health Administration), 135–139
overload, 62
oxidizers, 71

personal protective equipment (PPE),
    2–3, 32, 34–39, 48, 50, 67, 86, 94,
    100, 112
    Level A, 34–35
    Level B, 36
    Level C, 37
    Level D, 38
    use of, 38–39
personnel facilities, 15
physical hazards, 1–2, 6
physiology, 54
platforms, 13
point of operation, 82
powered industrial trucks, 101–115
    inspection of, 105
    safety programs for, 105–106
power tools, 75–79
protective clothing, 35–38
pullback mechanism, 23
pyrophoric, 71

radiation, 47–53
radio frequency, 49, 51–52

railway safety, 115–116
reasonable accommodation, 17–19
relief values, 75
resistance, 59
respiratory protection, 35–38
restraints, 23
retail services, 111–126
Reynaud's syndrome, 55
risk assessment, 5
ropes, 123–126
rotating wood stock, 83
rotator cuff syndrome, 54
rupture disk, 75

safeguarding, 20–27
safety evaluations, 134–142
Safety, Health, and Return-to-
    Employment (SHARE), 18
safety training, 14
safety valves, 75
sanders, 89
sanitation, 15–17
saws, 81
scaffolds, 12
sensing devices, 23
service facilities, 111–126
sheet metal shears, 89
ships, 117–118
skylights, 12
slings, 123–126
sprains, 54
sprinklers, 68
stair enclosures, 68
stairways, 14
standard threshold shift (STS),
    44
storage, 120–123
strobe effect, 83

tanks, 13
tendonitis, 54
tenosynovitis, 54
thermal hazards, 73–74
tinnitus, 40
transportation safety, 113–115

ultraviolet radiation, 97–98
underground utilities, 14
unfired pressure vessels, 74–76

vacuum breakers, 76
vapors, 82
vehicular equipment, 101–110
vibration, 82
violence in the workplace, 127–133
voltage, 59

walking surfaces, 10
walls, 13
warehouse facilities, 111–126
welding, 96–100
wood chips, 82
woodworking, 81–88
working surfaces, 10
workplace violence, 127–133
work practice controls, 85

About the Authors

Frank R. Spellman is assistant professor of environmental health at Old Dominion University. He is a professional member of the American Society of Safety Engineers, the Water Environment Federation, and the Institute of Hazardous Materials Managers. He is also a board-certified safety professional and board-certified hazardous materials manager with more than thirty-five years of experience in environmental science and engineering. He is author or coauthor of more than fifty books, including *The Handbook of Safety Engineering* (GI, 2009) and *In Defense of Science* (GI, 2010).

Revonna M. Bieber works for the Naval Medical Center Portsmouth in the field of industrial hygiene. Her work focuses on environmental health hazards and radiologic and healthcare safety. She is coauthor, with Frank R. Spellman, of *Occupational Safety and Health Simplified for the Chemical Industry* (GI, 2009) and *Environmental Health and Science Desk Reference* (GI, 2011).